REFLECTIONS OF A PRAGMATIC ECONOMIST

Emery Castle shortly after his return to Oregon State University in 1986.

Reflections of a Pragmatic Economist

MY INTELLECTUAL JOURNEY

Emery N. Castle

Oregon State University Press
*A co-publication with the Oregon State University
Department of Agricultural and Resource Economics*
Corvallis

All photographs are from the author's personal collection.

The paper in this book meets the guidelines for permanence and durability of the Committee on Production Guidelines for Book Longevity of the Council on Library Resources and the minimum requirements of the American National Standard for Permanence of Paper for Printed Library Materials Z39.48-1984.

Library of Congress Cataloging-in-Publication Data

Castle, Emery N.
  Reflections of a pragmatic economist : my intellectual journey / Emery N. Castle.
    p. cm.
  Includes bibliographical references and index.
  ISBN 978-0-87071-586-0 (alk. paper)
  1. Castle, Emery N. 2. Agricultural economists--United States--Biography. I. Title.
  HD1771.5.C37R44 2010
  338.1092--dc22
  [B]
                                                        2010019786

Oregon State University Press
121 The Valley Library
Corvallis OR 97331-4501
541-737-3166 • fax 541-737-3170
http://oregonstate.edu/dept/press

To Betty Rose and Cheryl Diana

CONTENTS

FOREWORD
by Daniel W. Bromley

Sidney James Castle, Josie May Tucker Castle, Ross Thompson, Fred Taylor, Merab Weber, Bill Alvey, Walt Bradley, Ed Bagley, Earl Heady, Kenneth Arrow, Ray Doll, Charles Warren, Marion Clawson, James Jensen, Gilbert White, Edward Mason, Charles Hitch, Edward Hand, Carl Stoltenberg, Roy Arnold, Thayne Dutson, Conrad Weiser, John Byrne, Curtis Mumford, Betty Rose Thompson Burchfield.

Our individual life history is the cumulative creation of subtle traces left by those whom, quite by accident, we encountered along the way. Nothing is written out for us. Life unfolds as a series of accidents. John Dewey insists that we enter life *in medias res*—as when one walks into the middle of a movie. Being prepared turns serendipity from a fickle friend into a constant companion. Emery Castle has never been caught by surprise.

To have known Emery Castle for 45 years is to have benefited from the great comfort of knowing a man of unrestrained modesty, unrivaled insight, and unequaled kindness. My own father died when I was 19—approximately the same age as Emery when he boarded that train in Wellington, Kansas to become a radio operator on 30 bombing missions over Germany. Life offers us a variety of ways in which serious growing up is urgently required. Some who have had the searing experience of war never seem to forget it—and too many seem unable to let the rest of us forget it either. The true spirit of a person is revealed in the personal memories they choose to convey to others. This is, after all, how we come to know others. I am quite sure that had Emery been able to provide his editor—and the reader—with a reason for skipping the years 1943-1945, he gladly would have omitted the entire affair. What horrible business to have been obliged to undertake; what anguish it must still incite.

On the subject of anguish, those of us from different pasts cannot comprehend the life of Sidney James and Josie May—just doing what parents have always done, only more of it. And from their story one comes to understand the tempered carbon steel inside Emery.

The unfolding of a life is a wonderful thing to behold—the decision points, the paths taken and avoided, the nudging help along the way, the careful deliberation of options, the consultation with others. What a joy to have the blanks filled in—graduate school, the Federal Reserve Bank of Kansas City, going to Oregon State University, stumbling onto roses, acquiring professional credibility, gaining skill as an academic administrator (imagine that), mentoring and inspiring many of us to become "like Emery" someday, guiding a respected Washington, D.C. research institute through perilous times, creating and leading an intellectual program to rediscover what is so marvelous about rural America, and then—as could be predicted—becoming the exquisite and consummate caregiver to his dearest Merab. It all fits, and it is so beautifully told.

But of course Emery has always been a teacher, and so here we see an example of the clarity of his thought process—and his marvelous ability to explain complex issues, and to connect them together in a coherent way:

> I favored a carbon tax on fossil-fuel use long before the climate-change controversy arose. I believe our economy and our lifestyles are much too dependent on an energy supply that has been available over time at a declining real cost, a use that assaults humans and the rest of the natural environment, directly and indirectly. I regret the nature of the climate-change controversy, because it is being waged on grounds that may damage perceptions of how scientific questions get resolved in science. Regardless of how it turns out on scientific merit, we may be distracted from thoughtful consideration of a more

fundamental issue—the interrelation of energy, economic growth, and the environment, with climate constituting only one part of the environment [chapter 7, p. 178].

The journey recounted here differs in such fundamental ways from the standard autobiography. It differs because it is Emery doing the telling. There are no boastful accounts of intellectual or bureaucratic triumphs. Emery never comes out on top. Emery brings people along. He explains, he teaches, he reasons, and he listens. He brings other people to his side. Many with whom he dealt, and this would be particularly the case during his decade in Washington, D.C., were accustomed to winning contests by enforcing the submission of others. Emery never considered human relations to be contests—they were, instead, teaching opportunities.

A few of us are blessed to have been in a life-long seminar with this wonderful man. Oh how fortunate we are. This son of Kansas reassures us that, indeed, there is nothing at all wrong with Kansas.

Daniel W. Bromley
Madison, February 17, 2010

A NOTE FROM THE CO-PUBLISHER
by Susan M. Capalbo

Because of Emery Castle's innumerable contributions to Oregon State University and the study of agricultural, rural, and resource economics, my colleagues and I in the department of Agricultural and Resource Economics welcome the opportunity to co-publish his autobiography, *Reflections of a Pragmatic Economist: My Intellectual Journey.*

What Dr. Castle (Emery to his friends, who are probably all of his acquaintances) understands, better than most, and what he has researched for his entire career, is the simple truth that rural and natural resources are intertwined and explicitly linked. In the quest to better human life and provide for a sustainable future, one needs to understand the nature and orientation of these linkages. In studying people and places, Emery contributes to pushing the boundaries of our profession—agricultural, rural, and resource economics—in unique and eloquent ways. He was one of the first proponents of a multi-disciplinary approach to rural-resource issues and has continued to champion the power of understanding collaborative connections. This is evident in the architecture of the Rural Studies Program at OSU, in his perspectives on the future of the land grant concept, and in his recent writing about an organizing framework to address the orientation of people to rural places: inside–outside orientation, outside–inside orientation, and inside–inside orientation, as described in Chapter 6. In addition to his leadership skills, he has promoted interdisciplinary graduate education at OSU and at the national level, developed the field of water research and the related valuation methods for instream flows, and developed a unique multidisciplinary perspective on people and places. During his long tenure at OSU, he has mentored and guided countless graduate students, many of whom have gone on to illustrious professional careers in the public and private sectors. He has truly shaped a generation of applied economists.

His commitment to excellence is displayed throughout the pages of this manuscript; his knack for providing a simple explanation to complex issues is manifested through the many anecdotes and circumstances he recalls during those early years at OSU and RFF in research, teaching, and administration positions. Emery has done a remarkable job writing an autobiography that unfolds so eloquently, as it combines the professional and intellectual contributions within his personal landscape. The signature of a great memoir is the ease with which the reader can connect the written word with the moment in time. On a personal level, *Reflections of a Pragmatic Economist* brings smiles and tears as I relate to Emery as a mentor, colleague, and friend; his autobiography serves as a reminder to me of the power of a committed person and a committed set of ideas.

Professor Susan M. Capalbo
Head, Department of Agricultural and Resource Economics
Oregon State University

PREFACE

This story is about the intellectual development of an applied economist who has spent more than half a century studying agricultural, resource, and rural economics. Both formal study and life experiences have influenced the point of view adopted in these studies and the methodological orientation that emerged. My objective for this book has been to determine and describe, as accurately as I can, the way these personal and professional experiences have meshed to affect the philosophical viewpoint I now hold. Economic methodology and philosophy of science treatises typically identify "right" answers, and numerous descriptions of "right" answers exist. I have made no attempt here to advance one more "right" answer. Rather, the effort here has been to provide an answer that satisfies one person—myself.

Economic analysts often make statements such as, "I feel comfortable with this approach." But what does that mean? Does it reflect the pronouncements of great philosophers, or eminent economists, or the predispositions of the analyst, or, possibly, all of the above? If predispositions play a role, where do they come from? In this book, I examine experiences that were distinctively intellectual as well as those that stemmed from my role in society.

I am the son of Kansas tenant farmers, and my first impressions were formed during the Drought and Depression years of the late 1920s and especially the 1930s. After World War II, I had the good fortune to obtain three academic degrees, thanks to the GI Bill of Rights. Largely for opportunistic reasons, these degrees and a great deal of my subsequent professional career were at land grant universities. In contrast, for ten years I served in a leadership position at a nationally recognized "think tank" that was created, influenced, and staffed by intellectually and socially elite people, including some of great wealth. Now that I have had time to reflect on these distinctive experiences, I know that the environment from which I came, together with formal study, interacted to influence both the professional choices and personal choices I have made. This is true especially of my use of economic theory and the public policy analysis I have conducted and influenced.

But why should anyone else care? There are at least two reasons. *First*, recognized or not, the prior personal experience of both theoretical and applied economists influences what they believe to be important in the policy applications they make and the positions they take in their professional work. Nevertheless, much research methodology literature presents and explains *the* way to do things; read literally, particular treatises often do not leave much room for choice. Yet, if an entire literature is considered, there is more room for choice than individual writings would suggest. Individual responsibility accompanies freedom of choice. Because there is room to maneuver, economists need to justify the choices they make rather than automatically follow a "mainstream" point of view existing at a particular time.

Second, the intellectual journey described has relevance for non-economists as well as economists. For more than twenty years, I served on boards and commissions as an appointee of five Oregon governors. Under the Oregon system, these boards and commissions stand between the state and federal legislatures and the agencies responsible for implementing the legislation they have created. The Oregon boards and commissions on which I served made rules, heard and decided appeals, and formulated legislative recommendations. During my service, I drew heavily on my economics background and my familiarity with research studies in my areas of specialization. Even though I seldom directly applied findings from particular economic studies, that knowledge made it possible for me to make unique contributions to our deliberations. For example, I served for seven years as vice chair of Oregon's Environmental Quality Commission with William "Bill" Wessinger as chair. Bill is a retired industry executive, an environmentalist, and a philanthropist. He is not a theorist; indeed, he has a healthy skepticism about conceptual approaches unless they are aligned closely with empirical evidence or realistic situations. He commented about the quality of my judgments, comparing them with those of his family lawyer, who he states is the wisest person he knows. The judgments of which he spoke so highly were usually based on deductions from, and applications of, economic models as well as reliable methods of acquiring knowledge (the philosophy of science and research methodology). Science and economics can illuminate many important public policy issues. And science and economics are too important to belong exclusively to people who work in those fields. Tool users, as well as those who fashion tools, were in my mind as I wrote the pages that follow.

The organization of the book follows closely the way my life has unfolded. Each chapter concludes with a section titled "Reflections." These reflections are my attempts to extract meaningful conclusions from a retrospective appraisal of experiences, including both successes and failures. The final chapter is, in part, a synthesis of these earlier reflections.

ACKNOWLEDGMENTS

In October 2005, a symposium in my honor was held at Oregon State University. Some of the people invited to speak were not familiar with my academic background except in a most general way. Prior to the event, I drafted several pages designed to explain how I happened to be at Oregon State early in my career, then at Resources for the Future for a decade before returning to Oregon State. Some of the participants either read the material, or at least looked at it, and expressed appreciation for it. Somewhat to my own surprise, I had enjoyed preparing the material. I suppose most people enjoy talking about themselves from time to time, and the same may apply to writing about oneself. In any case, it is appropriate to note that the symposium provided the original motivation for this book. A friend and colleague, Paul Barkley, read, commented on, and greatly improved that material, some of which was used in this book.

Some months after the symposium, I engaged in a telephone conversation with Dan Bromley, a former student and holder of an endowed chair at the University of Wisconsin. I told him I was considering writing an intellectual memoir that would include early life experiences. Dan immediately became enthusiastic, and within a week sent a quotation from Thorstein Veblen that may be found at the beginning of Chapter 8.

I then told my wife, Betty, what I was thinking of doing. She encouraged me as well. She may not have known that her access to our one home computer would be limited drastically for several months as a result of this new project. Nevertheless, she has been remarkably patient and understanding as I have labored. In the absence of her love, understanding, and cooperation, I doubt I could have found the energy or perseverance needed to complete the project.

I sent preliminary drafts of the first five chapters to my niece, Lana Castle, in Austin, Texas, for editing and word processing. Lana was a big help in getting me headed in the right direction. Lana is an author herself and provided important help early in the process. My friend and colleague, Bill Wilkins, economist and former dean of Liberal Arts at Oregon State, has been of great assistance. Originally he came on board to help with technical details in Chapter 2, which is concerned with my World War II experience. He ended up reading and commenting on the entire manuscript. I am deeply in his debt. Bill's wife, Caroline, also became interested in what I was doing and read and commented on Chapter 2 as well.

As soon as the first five chapters were in reasonably good shape, I shared them with Gwil Evans at Oregon State. Gwil is a marvelous writer. He and I have known one another for many years and once had cooperated on another manuscript I authored. I asked if he would consider becoming a kind of conservator of the manuscript if something should happen that would prevent me from finishing it. (At my age one thinks of such things.) I suggested he take a few days to look over the chapters and share his reactions. Gwil liked what he read and recommended that I proceed to publication as rapidly as possible. He also recommended I find someone to edit as I wrote and help put the manuscript in a form suitable for a submission to a press for review. I respect Gwil's judgment greatly. His advice massaged my ego and renewed my energy.

In addition to Bill Wilkins, other colleagues also have read and commented on the entire manuscript. Irving Hock is one such person. Irving and I came into the economics profession at about the same time and have had parallel experiences. He was at Resources for the Future for part of the time I was there. I sent him electronic versions of the various chapters as they came from the computer. He would then send me his reactions with observations on whether his experiences were similar or different from mine. Not only did Irving's observations improve the book, his sense of humor enlivened otherwise tedious hours at the computer.

My close colleague from the time I returned to OSU in 1986 until the present, Bruce Weber, also read and provided comments on the manuscript, especially those parts describing shared experiences. When Bruce speaks, I listen. His comments affected many of the pages in the book and especially those in chapters 6 and 7.

I am indebted as well to still another OSU colleague, Steven Buccola. Steve and I have had different applied economics specializations in our respective careers. We both are interested in the philosophy of science and how that relates to applied economics research. Steve is a rigorous thinker and an excellent teacher, researcher, and writer. I took my manuscript to him and asked that he review it rigorously and comment on anything that he thought would improve it. I also asked him to give particular attention to certain parts of the manuscript about which I had concerns. Steve took me seriously and prepared a page-by-page review. We then met, and he went over his review with me. I am grateful, and all my readers, present and potential, would be as well, if they knew everything I know.

Finally, but importantly indeed, I mention Cheryl McLean. Gwil Evans brought us together and recommended I ask Cheryl to work with me. She has edited the entire manuscript, but she has done much more than that. She is responsible for all of the illustrations in the book. She has done much to convince me to personalize the book by including descriptions of experiences in the text, and in endnotes and appendixes. She has taught me much about writing, but not enough to enable me to describe the quality of her service accurately and adequately.

Emery Castle
Corvallis, Oregon

1

MY EARLY LIFE
(1923–1942)

MY FATHER AND MOTHER

I begin by discussing, albeit briefly, my parents, Sidney James Castle and Josie May (Tucker) Castle. Each affected my life greatly, but differently.

Dad was born in 1884 in Ness County, Kansas, near Bazine, between Great Bend, Kansas, and Pueblo, Colorado. This is in the arid western third of Kansas. His father, Theodore Castle, died when my father was three years old. Not surprisingly, Dad knew more about his mother's family than his father's, as frontier family histories typically were preserved by oral rather than written means. My father's mother was Nina Swink, daughter of John Swink and Emma Baldwin. They were married in Pennsylvania, and both were descendents of early New England settlers. During the Civil War, Captain John Swink commanded a Union Army company called the Pennsylvania Sharpshooters.

Genealogical information about Mom is sparse. She was the sixth child in a family of seven. She also was born in western Kansas. Her father farmed and was an accomplished carpenter, but he was away from his family a great deal. My mother's mother was the only grandparent I have known, and I didn't know her well. She lived with us a few months

The Family of Sidney James and Josie May Castle. From left to right in the rear: their sons Robert Leroy Castle, Carl Theodore Castle, Emery Neal Castle, and Donald Ray Castle.

one year when she was of an advanced age. Her family name was Wright. The Wrights migrated to Kansas from Illinois, according to my mother. When asked about the ethnic background of her family, Mom would say, "English, Scotch-Irish, and Dutch."

As a product of his era and location on the Kansas frontier, Dad had only three years of formal schooling. He was the third child in a family of four. Because of the early death of his father, his mother reared four children born within a six-year time span. It is not clear how this single-parent household survived in the hostile frontier environment until the children became adults. No doubt neighbors assisted with the development of their homestead. Nina (Swink) Castle's fundamentalist Christian religion meant much to her. She read long scripture passages to her children almost daily. Her third child, my dad, always had a Bible in our house, but he did not spend much time reading it. He recalled his mother's scripture readings with little fondness. My family were not members of any church. I was inside a church for the first time when I was in high school, and the first church wedding I attended was my own.

Dad was the embodiment of the "Do unto others" ethic to a greater extent than anyone I have ever known. His words for it were: "Put yourself in the other feller's shoes." When I was a small boy (perhaps

10 years old), the local banker told Mom in my presence: "Sid Castle is as honest a man as I have ever known. He will always be in debt, and he will always pay his debts." Dad repeatedly told his four sons, "I cannot give you wealth, but if you work hard in this country, you can get a good education." That opportunity had not been open to him. To the best of my knowledge, he never curried favor with anyone unless he respected the person ethically. Conversely, he extended friendship to many who were not well accepted in communities where we lived. An example was a grocer from Syria who became a buyer and seller of live-stock and a friend of Dad's. I did not acquire religious or racial preju-dices from my nuclear family, for which I have long been thankful.

My father was a gentle man. He was firm as a father but seldom inflicted punishment. I tried to please him because of the respect I had for him, and I very much wanted him to be proud of me. He was the greatest moral influence I have had in my life. Even now, when faced with a moral choice, I often ask myself, "What would Sid Castle do?" With his behavior, he illuminated propositions and principles I later encountered in the formal study of ethics. When I discussed the wars in our nation's history that I was studying in school, he often gently pointed out differences in sacrifices among groups within society.

Mom and Dad were very different and complemented one another accordingly. She was exceedingly practical. Rigorous and tough in establishing the relative importance of objectives, she adjusted priori-ties readily when conditions changed. She had more formal education than Dad, having completed eight years of schooling. Her intuitive grasp of number manipulation was impressive indeed. If anyone ever tried to "short change" Mom, they undoubtedly regretted the attempt. A pocket calculator for her would have been completely redundant. My early professional interest in farm management originated, I am sure, from her demonstrated managerial capacity in the house and on the farm. So far as I can recall, Mom and Dad shared in major decisions easily and readily. When seeking permission to do something as a child or youth, I do not recall ever trying to play one against the other.

Mom was seldom idle. She worked inside the house and in the field, sewed, rode horses, milked cows, and, so far as I know, was good at every craft she tried to master. She was sharp of tongue and not tolerant of those less committed to work and family than she was. When he was in his fifties, Dad spent weeks in a Catholic hospital in Winfield, Kansas, as a result of a botched appendectomy. My parents did not have funds to pay the hospital bill when it came time for him to be discharged. Mom negotiated an agreement with the hospital to deliver dressed chickens to them until the bill was paid. This required several years, until near the end of Dad's life.

Clearly, my parents influenced me greatly, individually and together. Words would fail me if I were to attempt to describe the problems and challenges they faced and the efforts they expended. Their enormous, but quiet, sense of personal responsibility abides.

THE THIRD OF FOUR SONS

My parents were married in 1915 in Dighton, Kansas. They homesteaded in Grant County, Kansas, where my oldest brother, Robert, was born. They then moved to a second homestead in eastern Colorado, where they lived in a sod house dug into a hillside (called a dugout) and did not prosper. Their second son, Carl Theodore, was born there in 1917. (For a vivid account of conditions in that region and during those years, read *The Worst Hard Times* by Timothy Egan.) When Carl was quite young, they decided to move to eastern Kansas, where oil fields were being developed and where, they were told, steady work was to be had.

In 1920 they put their possessions in a covered wagon and spent the summer moving to Greenwood County, Kansas. Mom and my two older brothers were in the wagon while Dad drove the horses. The wagon was heavily loaded, and they stopped in the afternoon so the horses could graze and rest. At night Mom would prepare a meal on the stove, which was vented through the canvas wagon top. Cook stoves were important possessions for homesteaders. Typically, there was a chamber for firewood or other fuel, an oven that supplied even heat, and a cooking surface. These stoves were space heaters as well. The space-heater feature was far from a blessing during Kansas summers, when frontier women labored over hot stoves to prepare food for families and work parties.

I was born in Greenwood County in 1923. Dad was a "roustabout" in the oil fields. Mom came to provide a boarding house in the oil fields

Josie May and Sidney James Castle; his artificial left eye was the result of a water well-drilling accident at the age of 21.

at the request of Dad's fellow workers. Some of them then asked to have monthly Saturday night dance parties. Mom said yes, but she decreed there would be "no cussin', no fist fights, and no likker." Her Saturday night dances continued until my family left the oil fields three years later.

My family returned to farming in 1926, when I was three. They rented a farm on the east side of the Arkansas River in Cowley County, Kansas. This was river-bottom soil, and they produced vegetable crops as well as cereals. The farm was owned by a lawyer in Winfield, Kansas, who was impressed by the honesty and industry of his tenants. My younger brother, Donald Ray, was born there in 1929.

HARD TIMES IN THE GREAT PLAINS

The stock market crash occurred in 1929, with the full reach of the economic depression to follow. My parents were determined their four sons would have a better life than they had or would have. They explored farms for rent that would provide greater potential for growth than their small river-bottom farm. They then made, in retrospect, a poor decision. They rented a 320-acre farm in northern Cowley County near the small town of Burden. The farm was located in the foothills of the Flint Hills, which are marvelous limestone soil-based grazing lands.

Unfortunately, the farm my parents rented was in cultivation with soils not well suited to crop production.

Shortly after my parents agreed to rent the farm, they went to Winfield to tell the owner of the river-bottom farm that they would be moving before the next crop year. The owner told them he had purchased an upland farm with good soil for them because he had been pleased with their honesty and hard work. When telling and retelling this story, Mom would say in a matter-of-fact way: "Of course, we had given our word to the new landlord and could not back out."

The Depression, signaled by the stock market crash of 1929, became worse in 1930 and continued until World War II. And the Depression was accompanied by the Drought. The drought years of the 1930s in the Great Plains were unprecedented in our nation's history (USDA 1941). Tree-ring studies show that dry and wet years often occur in clusters. There is disagreement as to whether this occurs more than might be expected if one were sampling from a normal distribution of wet and dry years. In any event, farms were foreclosed, and people in rural towns became unemployed. Banks failed and bankers often committed suicide. Land owners with good farms had their pick of tenants. Even so, rent payments of one-third of crops were inadequate to cover taxes and mortgage payments for land purchased in the 1920s.

It was in this environment that I formed my first impression of rural America. During these years, two brothers, friends of my two older brothers, lived with us much of the time. They were from a family of nine that lived in a nearby town. Their father was unemployed. The two brothers worked on our farm in exchange for their food and lodging, although they were free to leave when they could get paid employment elsewhere. I recall awakening one night and overhearing a conversation between my parents. They were concerned we would not have enough home-canned food to last through the winter. They exclaimed to one another about the enormous appetites of the four young men who ate with us—my two brothers and their two friends. For the remainder of the year, we ate much unprocessed cereal and wild game—jack rabbits, cottontail rabbits, and squirrels. Although we were living in poverty, my family was providing help to a family worse off than we were. I was to remember this later when I learned that charitable giving, as a percentage of wealth and income, tends to decline as wealth and income increases.

When we moved near Burden in 1930, I was in the second grade and attended Grand Prairie, a one-room country school located two and one-half miles from our farmstead. Our music instruction was enhanced by an itinerant music instructor who visited us every two weeks or so with a scratchy phonograph. It was from that phonograph that I heard my first symphony. To graduate, we had to score 75 percent or better on a county-wide examination covering reading and writing, history, government, arithmetic, and geography. In some states in the Great Plains, one-room rural-school students were required to pass the county examinations to be admitted to high school. No such examination or requirement existed for city-school students. I harbor no complaints about the quality of my one-room schooling. When I was in the fifth or sixth grade, I began to get into mischief (tossing wet paper wads, passing notes, noisily passing gas, for example). After I was punished at home, I was assigned to recitation sessions with the eighth graders in history and geography and required to write a report about Canada. When I was in the eighth grade, my teacher permitted me to help with first- and second-grade kids studying arithmetic. Perhaps the highlight of my primary-school accomplishments was winning the Cowley County public speaking contest for all public-school students.

The term "allowance" for a child was not in my vocabulary at that time. The thought that, as a child, I might have some right to a portion of my family's income never entered my mind. We often went to Burden (population about 500) on Saturday nights. When I was 9 or 10, my parents often would give me a dime to spend as I wished. I always spent five cents for *Liberty* magazine and usually five cents for a candy bar. I would try to limit myself to reading a part of my magazine each day in the week. I usually failed to exert that much self-discipline and would read the entire magazine on Sunday, but reread it several times during the week. The lack of opportunities for intellectual stimulation in my part of rural America is a vivid memory, although I would not have described it in that way at that time. I read everything I could get my hands on, except the Bible. For reasons I do not understand, the Bible did not have great meaning for me.

When I was in grade school, for a period I became something of a loner. During the summers and on weekends, I often would hike into pasture land or other uncultivated places. Sometimes I would build a small fire and, if I had anything to read, would do so beside the fire. I always doused the fire because I knew about prairie fires. I have come to believe that Ross Thompson, a friend and neighbor, noted my behavior. Ross Thompson persuaded Dad that I should have a pony and supplied me with one. That was how I came to have Peanuts. Peanuts was a Texas cow pony who came to Kansas in a shipment of Texas cattle sent to the Flint Hills to graze that lush grass before being sent to market. Texas ranchers often said "thanks" with a saddle pony to those they contracted with to provide pasture for their cattle. Peanuts was a small, strawberry-roan pacer and an excellent cow pony. We were together several years. I rode him to school in the winter and together we herded cattle and swine in the summer. Yes, swine! I have been told it is impossible to herd swine from a horse. As a cow pony, Peanuts probably was insulted when we first did so because he was adept in working a herd of cattle. Swine are intelligent creatures, and when they learned they would be rewarded with food and water if they did as we wished, herding them became a piece of cake. One summer Sunday I was ill. Dad told my two older brothers to herd the pigs. Not a good decision! We spent the following week finding and bringing stray hogs home from around the neighborhood. What an ego boost for me!

HIGH SCHOOL

It was seven miles from our farmstead to the Burden High School, so transportation to and from high school was a problem. My older brothers rode horses, but fourteen miles a day was a long ride and required considerable effort from horses and riders. My parents tried something different for me. When I was a freshman, they arranged for two rooms with a hot plate, which I shared with another freshman. Our diet was inadequate, and I developed kidney problems stemming from malnutrition. I missed several weeks of school. When I was able to return to school, a neighbor girl and I drove the fourteen-mile round trip daily in an old car. Her family owned the car, and my family paid for gas and oil.

During the summer following my freshman year, the owner of a river-bottom farm west of the Arkansas River and north of the small town of Oxford, Kansas, came to our home. He raised the possibility of my parents becoming his tenants. He was a lawyer named Fred Taylor, who worked for the Federal Land Bank in Wichita, Kansas. His relatives became acquainted with my parents when they farmed east of the river. Fred Taylor's farm was nearly all orchard, divided between peaches and apples. Oxford was a progressive little town (population about 1,000) surrounded by diversified medium- and larger-size farms. Fred's visit caused great excitement and joy in my family; it did not take my parents long to accept Fred's offer to become their landlord.

I became a student at Oxford Rural High School during my sophomore year. It was a great place for me. It had a talented and dedicated teaching staff. I flourished in all of my sophomore-year classes. Near the end of that year, the school principal called me to his office and personally designed my junior-year program. He included advanced algebra among the courses I was to take. Near the end of the following summer, before the start of my junior year, I contracted typhoid fever and missed the first two months of school. My principal came to see me during my illness and brought class materials when he did so. He did not believe I should attempt advanced algebra. My parents and teachers were concerned about my survival, but I did not believe I was going to die.

Disaster came to the family farm during my junior year. Dad and Mom had gone into debt to get the machinery needed to tend the apple and peach orchard on the river-bottom farm. For two years the farm was modestly successful. Then the Arkansas River went on a rampage. Water covered the orchard for more than a week. When the water subsided, several inches of silt were left behind. The orchard was destroyed and had to be removed. Dad never recovered, emotionally or financially.

Despite these difficulties, I enjoyed and did well in school. Several teachers made indelible impressions. One, Merab Weber, became my wife many years later. She taught English grammar, English literature, speech, and dramatics. Extra-credit work was required if we wished to be considered for an A or B grade. This was not popular with some mothers in the community, but her principal made it stick. She was a demanding teacher and taught all students with both firmness and love. I am indebted to her for the fundamentals of grammar and for teaching me almost everything I know about public speaking. (There is more about Merab in various places in this book, in particular in Appendix A.)

Bill Alvey was my vocational agriculture teacher. He was not rehired for the school year following my senior year. Bill almost defied description. He was short and skinny, with red hair and blazing blue eyes. Bill had been an honor student at Kansas State, but he always seemed disorganized and slightly confused. He had a million ideas every waking moment of his life. Many were unrealistic nonsense, but some were imaginative and valuable. He was not well accepted in the Oxford community, and some of the other teachers regarded him with disdain. It was taken for granted that farm boys would take vocational agriculture. I was not much interested in, nor very good with, subjects such as farm mechanics and other vocational endeavors. Nevertheless, I found myself in a vocational agriculture class taught by Alvey's predecessor, Mr. Lowe, soon after I enrolled in Oxford High. Bill Alvey came the following year and attached himself to me from that time until his death more than sixty years later. He had flirted with socialism and left-wing activities when he was in college. Even so, he encouraged his students, with help from their parents, to acquire crop and animal enterprises as a way to accumulate wealth in a capitalistic economy. He became an avid and strident Democrat after he left teaching and began to farm.

Years later, when Jimmy Carter became President, Bill thought I should be Carter's Secretary of Agriculture. It didn't matter that I was not interested, nor that there was not even a remote chance the Carter people would be interested in me. At that time, Bill was farming in Colorado and active in politics there. (Bill probably never learned that Bob Bergland, Jimmy Carter's actual Secretary of Agriculture, and I became friends. We met when he was Secretary and later served jointly on committees and study groups.) Bill Alvey was a friend to the end of his life. He contributed something original to every issue that came before him.

At the beginning of my senior year, my 35 classmates elected me senior class president. As such, I participated in numerous community activities and did a modest amount of public speaking. The Oxford Rural High School faculty selected me to receive the American Legion Award given to the outstanding senior each year, although I was not at the top of my class scholastically. I graduated in 1941 without firm plans as to what I would do next, although I had ambitions to attend a four-year college or university.

BRADLEY'S STORE

Shortly after my graduation from high school in the spring of 1941, I came home one day to find everyone in a state of excitement. Walt Bradley, owner of Bradley's Dry Goods Store in Oxford, had paid a visit. He had decided to open a branch store in Wichita, home of the Boeing Aircraft factory. The United States had not yet entered World War II, but the war had stimulated economic activity in Kansas, especially at Boeing Wichita. Mr. Bradley discussed with my parents the possibility that I might manage his local store while he launched his Wichita branch store.

Walt Bradley and I did not know one another at that time. He had gone to the principal of the high school for names of recent graduates who might manage his local store. Apparently, when he spoke with my parents, I was the only person he had considered seriously. Of course, by most standards it was not that big a deal. Oxford was not a large place. The store stocked quality work clothes and shoes for men, house dresses for women, underwear for everybody, and other assorted items. Mr. Bradley had left farming some years earlier and invested in a pool hall in Oxford. From the first, he sold a few items of merchandise in the pool hall. The merchandise sales won the competition for space, and the pool hall disappeared.

When Mr. Bradley called on my parents, I had not abandoned the possibility of entering a college or university that fall. However, my financial resources were almost nonexistent. Bill Alvey had submitted an unsuccessful Sears scholarship application for me to attend Kansas State. (I always suspected he missed the deadline.) Although I knew nothing about working in a store, I recognized I could live at home and accumulate a small capital stake if I did so. I accepted Mr. Bradley's offer.

Mr. Bradley proposed that I work on salary and commission. We agreed I would work for the summer before we made an agreement for an entire year. I soon came to appreciate this laconic 60-year-old Kansan. I do not believe he had completed more than eight years of formal schooling. We got along famously. He told me later he had doubts early in my employment that I would ever learn to wrap a package properly. After I organized his financial records consistent with double-entry accounting, he said he was thankful they did not teach that subject when he was in school. I judged he was grateful for the service rendered but was glad he did not have to do it.

As a farm boy, I thought people who worked in town probably did not work very hard. I learned differently. I organized the merchandise, dressed windows, replenished stocks, and kept the books. I came to enjoy the work, but only after I learned how to communicate with women when they were considering house dresses or unmentionable apparel. The store hours were 7 a.m. to 6 p.m. five days a week, and 7 a.m. to 9 p.m. on Saturday. By 9 p.m. Saturday I was exhausted! Even so, on weekdays there were periods when I was the only person in the store. Still, I enrolled in two correspondence courses from Kansas State: Principles of Economics and History of Breeds. I received five hours of college credit, little understanding of economics, and the conviction that I did not want to major in animal husbandry if I were to attend Kansas State.

And then came Pearl Harbor Day. On that December Sunday, three of my former classmates and I had gone for a ride in the countryside. We returned to my home in the afternoon and were listening to the car radio when the news was broadcast. All of us, men and women, knew our lives had been changed. We were serious and sad.

In late summer of 1942, Mom and Dad took the only vacation of their lives. Mrs. Bradley took over my duties at the store and I, together with my younger brother (then 13), accompanied them. We were away more than two weeks but not three. The first day we drove from Oxford, Kansas, to Dighton, Kansas, where my widowed Aunt Nettie McCoy lived. From there we went near Two Buttes, Colorado, where Uncle Emery and Aunt Addie Castle resided. It had been several years since either of my parents had seen these siblings.

Aunt Nettie had acquired sole ownership of a hotel in Dighton when her husband died some years earlier. The hotel had been built during the more prosperous 1920s and at one time was considered to be one of the nicer places to lodge in western Kansas. It had suffered

Aunt Nettie's Hotel, Dighton, Kansas.

the ravages of the Depression by 1942, but Aunt Nettie kept it clean and in reasonably good repair. The evening we arrived, Mom wasted little time in asking, "Nettie, how long has it been since you have seen a movie?" Aunt Nettie replied that it had been a long time because she could not afford a desk clerk at the hotel in the evenings. Mom said, "Sid and I want to take you to the movies tonight. You can tell Emery what he needs to do to be desk clerk." That was the first I knew of this plan, but I decided that if I could run a store, I could be a desk clerk. Aunt Nettie took some persuasion, but eventually they all left for the movies except me. As she left, Aunt Nettie said she really did not think anybody would request a room.

Soon after they left, a short, white-haired man, followed by a female person of almost the same description, came through the front door, approached the desk, and asked, "Where is Nettie?" By that time, I began to fear I was in over my head. I did not know much about the different rooms available, nor just what I was supposed to do other than have them register their names. The white-haired man said they had stayed in the Dighton Hotel several times over the years. He said they lived in Emporia, Kansas, and Dighton was on the route they traveled to a cabin they owned in Colorado. He then completed the registration form and pushed it across the counter to me. It read: Mr. and Mrs. William Allen White, Emporia, Kansas. I was sufficiently well informed to know I was in the presence of the most famous Kansan alive. I was filled with confusion and deeply embarrassed by my clumsiness as I carried their bags to their room. When Aunt Nettie returned I told her, somewhat breathlessly, who she had as guests. She said, in a matter-of-fact way, "They stop here a lot."

I remained in Bradley's store through the winter of 1941–42, and the following spring and summer. By then I knew I did not want to remain in the store beyond that year. I considered enlisting, but Dad discouraged that course of action. I then decided that I would start college in the fall of 1942 and remain there until I was drafted. Bill Alvey submitted a second Sears scholarship application. This one was successful and provided further incentive for me to enroll at Kansas State.

KANSAS STATE

Bill Alvey had prepared me well. I knew in advance the agriculture curriculum at K State would emphasize the application of many sciences to agriculture. One course carried the title Freshman Lectures. It was taught by the head of the Psychology Department and was designed to equip students for learning in a college environment. Both then and in retrospect, I considered it an excellent course. We took numerous aptitude tests to provide information about subjects and careers we might wish to pursue. Near the end of the course, I received a sealed and stamped letter from the instructor. He wrote that my aptitude test scores indicated there was a high probability I would attend graduate school later. He went on to recommend courses I should take to prepare for graduate work. I thought his letter was hilarious and showed it to several people for laughs. I thought of myself as a slightly better-than-average student who was not much interested in basic science. Public policy issues interested me, but I did not know how to pursue them as a field of study. At that time, I believed the high schools at Topeka, Wichita, Great Bend, and other cities were much superior to little old Oxford Rural High School. If I did not get all A's at Oxford, what chance would I have in graduate school?

By the fall of 1942, war mobilization was proceeding rapidly. My male classmates and I knew we would be in the armed forces soon if we did not have physical disabilities. After one semester at K State, I returned to my home in January 1943 to await a call into the armed services. I was a 19-year-old naïve and immature lad.

THE ARMED SERVICES

When I returned to my home in January 1943, I knew it was just a matter of time before I would be drafted. I considered volunteering, but my parents, especially Dad, did not encourage that move. My attitude toward military service was ambivalent. I did not have concerns about my nation participating in WW II comparable to the questions that college-age students subsequently raised about the Vietnam and Iraq wars. I thought Hitler had to be stopped and also believed the United States had a responsibility to help in the stopping. Even so, I had unanswered questions about the violence that stemmed from wars. I was not sure that I could kill somebody if I was expected to try. On the other hand, others my age were marching off to war in great numbers. I was far from immune to the peer pressure that created. My decision to be drafted, rather than volunteer, was consistent with the ambivalence I felt.

A decision was required regarding the branch of service for which I would express preference. My three brothers all subsequently chose the Navy. I began my decision process with the self-evident proposition that the Armed Forces were dangerous places to be in wartime. I then asked myself what manner of death would be most likely in the different services. Death by drowning was unattractive, so the Navy was eliminated. I associated Army service with trench warfare duty, and I did not want to die in those miserable conditions. That left the Army Air Force as my first preference. To be sure, being shot from the air was not appealing, but parachutes existed. If a parachute did not open, death would not be prolonged. It was on this basis my choice was made.

It is frequently alleged that "Americans were all in WW II together." I believe there was truth to the observation. Everyone was expected to contribute in some way. If not, disapproval was typically expressed, and sometimes in not very nice ways. After World War II, I was glad I'd had a visible role in the war effort. Yet to this day I am ambivalent about wars. I consider wars to be the ultimate failure of a civilization. As an old man, I observe other old men and women sending young men and women to war, and I believe that to be wrong. I observe some acquiring great riches in war, even as others make the ultimate sacrifice. Yet I also believe the moral basis for all wars is not the same. My deep concern is that the incentives for war and peace are not distributed randomly across populations. It is, therefore, difficult to thoroughly explore nonviolent alternatives to war, even in democratic societies.

My World War II service was unlike any experience I have had before or since. My three war years—1943, 1944, and 1945—expanded my horizons greatly, but when they were over, I attempted to put those experiences behind me. My daughter has said I talked little about them during her childhood and youth. I have made an effort in Chapter 2 to be more forthcoming.

REFLECTIONS

I retain three overriding impressions from my life as it was until I entered the Armed Forces at 19 years of age. One impression pertains to my parents. Another involves rural education. The third concerns public policy, especially agricultural and rural policy—how those policies were regarded then and what they have since become.

MY PARENTS: Two emotions struggle for dominance when I think of my parents. One is sadness; the other is gratitude. The sadness occurs when I think of the struggle they waged to survive and rear a family in an exceedingly hostile environment with small material reward. Until World War II, the economy that provided a market for the goods and services my parents sold underperformed significantly. Over the course of their lives, they lived in poverty and accumulated little; they worked long and hard, were honest, and set admirable examples of self-reliance for their four sons.

The gratitude I feel is because of the example they set despite the environment in which they lived. As I came to adulthood, I formed attitudes about individual responsibility and education that have served me well. There were few cultural endowments and little intellectual stimulation in rural Kansas where I was reared. Yet there was much less of the racial and ethnic prejudice that has since afflicted so much of rural, as well as urban, America. To my everlasting benefit, my parents imbued me with the belief that I should make every possible effort to improve myself with whatever educational opportunity might be available.

RURAL EDUCATION: Schooling and education have been powerful forces in the transformation of rural America, especially during my lifetime. James Coleman, eminent sociologist, was to write that social capital formation necessarily had to precede schooling as a means of human capital formation.[1] Parents in rural America apparently grasped intuitively the truth of this astute observation long before Coleman articulated it. Private education was not unknown in rural places, but if every person was to have a chance to become schooled, community cooperation was necessary. Land was set aside for schools, and opportunities were created for bright young women to prepare themselves to teach in those school houses. One can quibble about the details, but, in fact, hundreds of thousands of young rural Americans have migrated to urban places during the twentieth century. Rural education can rightly take much credit for this remarkable accomplishment. I return to this subject later in this book.

I have received enormous benefit from rural public education. Even though I had no such aspirations when I graduated from high school, I have spent much of my professional life concerned with this subject. When I finished high school, my principal ambition was to acquire a four-year college or university degree. The people I knew who had had such an experience seemed somehow different than those who had not, although I could not have stated those differences with precision. Perhaps a greater breadth of perception was revealed as they spoke. Or perhaps it was the way they spoke. In any case, I wanted to be like them. No one in my family had gone to, much less graduated from, a four-year college or university. By then I had read about history and current events. I had vague thoughts about being a participant in important matters, and I believed that would be more likely if I had more education. My parents were supportive but were at a loss to provide advice or financial assistance.

As I reflect on my schooling to that time, the question arises as to how well prepared I was to seek fame and fortune outside of Sumner County, Kansas. Given the circumstances that existed then, I have little criticism of the one-room country school that I attended through the eighth grade.[2] Neither do I fault Oxford Rural High School generally. The instruction was not equally good in all subjects, of course.

The teaching of mathematics was not outstanding, unfortunately for those students who would attend colleges or universities in search of a degree in the sciences or engineering. Most of my classmates intended to remain in the community as farmers or workers in local business enterprises. Many of the young women intended to enter nursing or do clerical or secretarial work. In my class of 36, only a limited number aspired to qualify for a profession by attending college. In my case, only Bill Alvey provided concrete, specific advice. He *knew* I should study agriculture at Kansas State.

With the benefit of hindsight, I now know it was my female classmates who were the most handicapped. They had many fewer opportunities to achieve their potential than did those of my gender. There were only a few professions they appeared to consider seriously; teaching and nursing were among those mentioned most frequently. Nothing I recall prepared us for the enormous change in the role of women in our society that was to occur soon. It was World War II that set the stage for the women's movement that was to come later. Most of the women my age were born too soon to benefit directly, although all of us have benefited indirectly, of course. Our economy, as well as others, became much more vibrant because women entered the workforce in record numbers during and after World War II. I consider the feminist movement to have been the most significant social change to occur in the United States in my lifetime. As far as I am concerned, it was long overdue. My mother could do most of the things my dad or her sons could do, and often better.

The feminist movement, of course, can be considered as a part of the civil rights effort generally. My comments about greater female equality are not intended to detract from the significant and highly desirable progress that has been made toward greater racial equality.

PUBLIC POLICIES FOR AGRICULTURE AND RURAL AMERICA: Franklin Roosevelt was elected President in 1932, and the New Deal was born. Even then, I was conscious that something important was happening in the society around me. My dad had been a lifelong Republican but, along with many Kansas farmers, became a Democrat at that time. Even though I was only 9 years of age, I sensed the desperation as Kansas farm and rural people struggled. The farm policies that came into existence after 1932 provided a few rays of hope for these desperate people.

A few years ago, I visited the FDR memorial in the District of Columbia for the first time. There are four sections in the memorial, one for each of FDR's four terms as President. In the first section is to be found a man in old clothes seated on a stool outside the door of a shack. He is listening to a small radio that is broadcasting the words of an FDR fireside address. I experienced a significant emotional reaction when I viewed that exhibit for the first time. It was Dad who was seated on that stool! He seldom failed to listen to an FDR fireside chat. Perhaps that resonant voice coming over the airways provided small rays of hope.

The early New Deal farm policies were a grab bag of programs designed to provide farmers with some financial relief. I simply cannot imagine a valid objection to the meager payments my parents received occasionally. Those payments presumably were to compensate for conserving the soil or were related, in some way, to the amount of produce taken to the market. More realistically, their purpose was to provide impoverished people with a modicum of purchasing power.

The above situation is in contrast to what agricultural public policies have become. These policies and programs have been captured by the agricultural-industrial complex, sometimes labeled agri-business. These special interests moved to identify and associate with agricultural public policies and programs. Obscene subsidy payments and wasteful ethanol programs are the results. Every five years or so, Congress reviews and modifies the "Farm Bill." These farm bills are often thought to constitute rural public policy. In fact, most rural people are not affected, and many important rural problems are not addressed by farm bills. There are clear political advantages in tying farm policies to acres of land and bushels of commodities, but by doing so, many important issues—for example, rural poverty, rural health, and rural education— are bypassed. Great mischief results from treating "agriculture" and "rural" as synonymous entities.

Agriculture is a dynamic industry that changes over time and may, or may not, be the most important economic force in a rural countryside. The process by which an agricultural industry becomes modernized and then fully integrated within an urban society is not well understood even today. Although there has been experience with this process in many countries around the globe, troublesome problems reoccur. Early in the process of agricultural modernization, farming typically is a labor-intensive industry. As modernization occurs, farm labor is displaced. Unless urban employment is available, or unless rural employment opportunities are created, rural underemployment, rural low income, and rural poverty become by-products of agricultural modernization.

Adam Smith's *An Inquiry into the Nature and Causes of the Wealth of Nations*, written in 1776, is often the starting point when economists consider the economics of growth and change. It is inevitable that some important considerations will be neglected in a theory of such scope. Smith noted that the scope and size of available markets were closely related to the growth of firms and the development of industries. Farms and farming are not exceptions, except that early in the process of economic growth of a nation, farms are numerous and typically scattered over great distances. Other industries often can change locations to be close to consumers or raw materials. Farming is tied to the land. Farms grow and consolidate, to be sure, but at a less rapid pace than many other industries. This means that excess farm labor usually must migrate if it wishes to participate in the fruits of economic progress. Rural young people and their parents often know this well: witness the hundreds of thousands of young people who have left the countryside. Yet there are many others who have not migrated. Economists sometimes attribute this failure to migrate to a resistance of rural people to change. Those who make such allegations should consider the effort and sacrifice these people made to get to their rural homes in the first place, in the hope their future and that of their families would be better. Space, distance, urban economies, and population density are the economic welfare dimensions of agricultural and rural economics (Castle 1991).

2

WORLD WAR II
(1943–1945)

PREPARING FOR BATTLE

It was an unseasonably cold and windy day in April 1943 when Mom and Dad accompanied me to Sumner County's Selective Service office in Wellington, located about 15 miles from our home. There, orders were waiting requiring that I report to Fort Leavenworth, Kansas, for induction into the armed services. I do not recall anyone else boarding the train in Wellington that day. I do recall, in vivid detail, my parents, in their clean, worn, and faded clothes, waving goodbye against the bleak Kansas backdrop. The United States had been a declared participant in World War II for only a little more than a year. My parents obviously feared I would soon be engaged in combat.[1] And, of course, it was an emotional time for me. I sensed my life was changing, but in ways I could not imagine.

After induction at Fort Leavenworth, I took aptitude tests and was assigned. To my great surprise, the tests were intellectually stimulating and enjoyable. I had no idea how they would be used, but I believed I had done well. The Air Force had not been separated from the Army at that time, but my stated preference for Air Force service was realized. I became a "Wild Blue Yonder Guy."

I was sent to Jefferson Barracks near St. Louis, Missouri, for basic training. We were told our basic training would be no different from that taken by Army infantry recruits, similar to the experience of thousands of others going into the Armed Forces at that time. I recall Thursday afternoon inspections with little fondness. We would march in full uniform, sometimes with a rifle, to the parade grounds. It was summer 1943 before we completed basic training, and outside temperatures often were high. Between two and three hours were required to permit each company to pass before the reviewing stand. When we were not marching, we stood at "parade rest." This was more than many human bodies could endure for long periods, and numerous soldiers collapsed. I could hear the noises related to their fall as they went to the ground. Of course, we were not permitted to offer assistance. Rather, we would wait for the "meat wagons" to haul them away. I judge healthy young men and women were not damaged by such collapses, and I never had one.

Dad came to see me when I was in Jefferson Barracks. He traveled by bus across half of Kansas and the full width of Missouri. He rented a room in St. Louis, and I went there to see him. His visit was most enjoyable, but he seemed much older than when I had told him goodbye a few months earlier. Mother told me later she had urged him to travel to see me because he had become despondent after I was inducted.[2]

When I completed basic training I was told I had been classified as a member of the Army Specialized Training Program and would return to Lincoln for additional testing and assignment. I had never heard of the program and had no basis for forming expectations. After arriving at Lincoln, we were given many competency tests and were told the test results would determine to what university we would go for "specialized training." I was asked to indicate a preference between engineering and military government. I do not recall that we were given much detail about what either would entail in practice. In any case, I chose military government because of my interest in social problems and public policy. When my test results were interpreted for me, I was told my aptitude for mathematics was superior to that for languages and that I had been assigned to study engineering rather than military government. My image of army engineering was building bridges in a swamp, or some similar place. I asked what my fate would be if I refused the engineering option. The reply was that I probably would be trained for a combat assignment. By that time, my ambivalence concerning the war had largely disappeared. I reflected that combat was the reason I was in uniform, and I rejected the opportunity to study engineering. A questionable decision by an immature soldier!

I was then returned to Jefferson Barracks for reassignment. I was alternately given guard duty and kitchen police (KP) assignments. I was able to write to my family, but I was on duty during most "mail calls" and received little mail. I began to doubt I would ever make a significant contribution to the war effort, and to believe I would have the rank of buck private forever. This was the absolute low point of my military career.

Then things changed. I was given orders to report to the Sioux Falls (South Dakota) Army Air Base to train to be an aircraft radio operator. The training was interesting but not difficult. I believe (based on my recollection) the course was designed to be completed in twelve weeks. We were rotated among shifts—two weeks daylight, two weeks swing shift, two weeks graveyard. There were no chairs in the graveyard-shift classroom, to prevent students from falling asleep. I had pneumonia while at Sioux Falls and lost one week of school. Never mind, I just fell in step with the class following the one to which I originally was assigned. My first exposure to sulfa drugs occurred when I was in the hospital. They worked well, or I assume they did, because I recovered quickly.

From Sioux Falls, I went to Yuma, Arizona, for aerial gunnery training. It was early spring and the weather was a welcome change from the South Dakota winter. We were assigned to sleep in tents near the flight lines. I recall the airplane engines began to roar about 3 A.M. Much of the training made use of shotguns that were fired at clay pigeons. One exercise required we stand in the rear of a pickup truck firing a shotgun at clay pigeons hurled toward us from 90-, 45-, and 22.5-degree angles as we proceeded around an oblong track at about 10 miles per hour. The trick was to aim behind the pigeons, rather than in front of them, when firing the shotgun. To lead them was the intuitive, but wrong, thing to do. The analogy was to flying aircraft. Envision two flying aircraft, one the attacker, the other defending. By aiming behind the attacking plane, we compensated for the forward movement of the aircraft (pickup truck) from which we fired.

After Yuma, it was off to Dyersburg, Tennessee, with a brief stop at Lincoln, Nebraska, where I met the crew with whom I was to enter combat. We were designated a B-17 crew, and that was the aircraft we flew for the duration. At Dyersburg, I came to realize that combat was imminent. This same realization had come to others, and our participation in the various exercises was intense and serious. Also at Dyersburg, I became acutely conscious of the officer–enlisted soldier distinction. We got to know our officers personally and well, but it was clear they had many privileges we did not enjoy. We were a crew of 10 men. The officers were pilot, copilot, navigator, and bombardier. Enlisted men were crew chief and top turret gunner, radio operator, two waist gunners, ball turret gunner, and tail gunner. At some point during training, I was promoted to corporal, and for most of my combat missions I was technical sergeant. Our pilot was 25 years of age and the oldest among us. The copilot was the youngest, at 20. I was 21 when we finished our training.

During our time at Dyersburg, I went to Memphis several times with members of our crew. This gave me a glimpse of a culture different from that of the Midwest, the only other place I knew much about at that time. The dances and other social events we attended were all-white affairs, although black people were much in evidence elsewhere. While we were at Dyersburg, a terrible accident occurred that disturbed me greatly. A black man and his wife lived in a shack near the air-to-ground gunnery range where we practiced. One night, as they were sleeping in their shack, their bodies were riddled with 50-caliber bullets from an overhead B-17. Our crew did not fly the night this occurred, so we were not suspected of being responsible, but I was disturbed deeply by the attitude of my superiors. The tone of the group meeting where we received a reprimand could best be described as "naughty, naughty." I saw no mention of the event in the local papers. If the offending crew was identified and disciplined, I never learned of it.

After completing our training at Dyersburg, we returned to Lincoln, where our crew was assigned to a heavily loaded B-17 that we flew to

The Wild Blue Yonder Guy

England. We refueled and spent a night in the northeast, then continued over the North Atlantic to our British destination.

THIRTY MISSIONS

The missions in which I participated are listed at the end of this chapter. As I look at that list, numerous emotions and recollections compete for attention. I do not know why I have not discussed those missions more with family, friends, and others. It is not because of an "aw, shucks" false modesty. Nor is it because I have no pride in what we did. It is important to understand that those experiences are unlike any I have had before or since. When one does not have a point of reference, it is difficult to provide an accurate or realistic account of an event that has occurred. If one has no wish to glorify such an experience, it is easier to remain silent than to struggle to describe it. I have read descriptions of bombing missions that, to me, lack authenticity. This is not necessarily because of dishonesty or an intention to deceive. I believe much of the lack of authenticity stems from how difficult it is to describe an unreal event in a realistic way. There are notable exceptions. One such exception is the movie *The Memphis Belle*. Another is Andy Rooney's *My War*. Rooney does not write about bombing missions at length; most of his book is about other arenas. He did go on some missions as an observer and, in my opinion, provides observations that are realistic to a fault. Robert Grilley's *Return from Berlin: The Eye of a Navigator* belongs in this grouping as well. A recent addition to these accounts is Edward M. Sion's *Through Blue Skies to Hell: America's "Bloody 100th" in the Air War over Germany.* This book is woven around the experiences of Richard R. Ayesh, a B-17 navigator and Sion's uncle, who was assigned to the 100th Bomb Group on the same day I was. Ayesh kept a diary of each mission he flew, and these accounts are included in Sion's book. Ten of the missions he flew were also missions I flew. These accounts permitted me to check my memory on specific issues. His diary and my memory have proved to be remarkably consistent. Sion is a professor of astronomy and astrophysics at Villanova University. His book is far more than an account of his uncle's missions. He explores various facets of the World War II air war in Europe in great depth. I cite his book frequently in this chapter.

Before describing particular missions, I will provide a brief account of the military organization that was employed. Consider first the bomb groups that were scattered among numerous rural places in England. During the time I was flying (1944–1945), a fully constituted bomb group in flight usually consisted of 36 planes, although at times there were 37 planes in a group. When I began to fly missions, three squadrons of 12 planes each constituted a group. Later, this was changed to four squadrons of nine planes per squadron. Three bomb groups combined to form a wing. A combination of wings in England at that time was often referred to as the Eighth Air Force. That was inaccurate because the entire Eighth Air Force in England included B-17 bombers, B-24 bombers, and fighter planes. Nevertheless, I flew only on B-17s, and the squadrons, bomb groups, and wings in which I flew consisted only of B-17s. B-17s developed the reputation of being more durable than B-24s, although B-24s had greater range, speed, and carrying capacity.

We flew daylight bombing missions. We flew missions at night in training but never in combat. The Royal Air Force (RAF) did so, constituting a division of labor. By today's standards, most of our targets were not far away. Nevertheless, many hours were required from the beginning to the end of a mission. Typically we were given a 24-hour alert, enabling us to get "sack time" before a mission. On mission days we typically rose at two, three, or four o'clock in the morning. After breakfast mess, we were taken to a large briefing room where crew members assembled. The dramatic moment came when the briefing officer walked on stage and pulled the cover from a map to reveal our target. If it was a "milk run," there would be expressions of relief. Groans and gasps would signal the probability of heavy flak, fighters, or both.

After the initial briefing, we would go to specialized briefings—pilots to one, and navigators, bombardiers, and radio operators to other individual briefings. After briefing, we would be taken to our planes. Those of us with gunnery responsibilities would cleanse the 50-caliber machine guns of grease and install them in the plane. During gunnery training, we were required to learn to assemble the many parts of a machine gun blindfolded, a requirement I thought unnecessary. The first time we took off for a combat mission in darkness, I realized how wrong I had been.

The B-17s would take off one by one. This was followed by an operation so intricate that, even with nearly 70 years of perspective, I still think of it with awe. The B-17s would lumber down the runway, heavily laden with bombs and fuel, climb gradually, and then, on a typical English morning, disappear in the clouds. Other B-17s would be in the clouds ahead of us and still others would follow. We knew this same operation was going on all over England. We would climb through clouds for what seemed an interminable period and then break out of the clouds. We could then look around and see B-17s in all directions. Some had broken out of the clouds before we did; others came later. We assembled by squadrons, groups, and wings. The Eighth Air Force, aloft, usually would begin to move toward its target as the assembly continued. It usually was daylight by the time we were above the clouds, and, if so, we could see great distances. Watching those planes coming into formations above the clouds was one of the most beautiful and moving sights I have ever experienced. There were, of course, mid-air collisions, although I never knew anyone involved in such an accident.[3]

Even though the assembly of planes took considerable time, the journey to the target seemed to take forever. Actually, the two were closely related because some assembly could occur as we moved toward the target. Some crew members had little to do. After we became a lead crew, I sometimes had responsibilities as a lead radio operator, although messages were seldom transmitted on the way to a target. Responsibilities and tasks were welcomed because my attention would be diverted from what might happen later. When collections of slow-flying B-17s were headed in a particular direction, the German opposition had time to organize a defense. The closer we came to the target, the more certain they would become about where we would drop bombs. This is where tactics became important. The two main opposition weapons were fighter planes and anti-aircraft gunnery.

We feared fighter planes. They carried considerable armament and could inflict great damage on bombers in a very short time. Bombers could inflict damage in return but had limited maneuverability and were highly vulnerable to fighter attacks. The Germans established priorities among sites they wished to protect. High-priority sites were defended fiercely, both with anti-aircraft fire and sometimes with fighter planes.

After D-day in June 1944, Allied fighter planes dominated the Luftwaffe. The Allied forces had more planes, and fuel was a problem for Germany. We disrupted oil refineries with our bombing, so fuel availability limited the number of planes they could put in the air, their ground transportation, and the training of pilots. The missions on which I flew began in September 1944 and lasted until V-E Day in April 1945. Our fighter protection usually was quite effective. We believed our escort was "up there" someplace, even when they were not visible to us. Even so, if the Luftwaffe could strike at the right time, they could disable bombers and force lead crews off course. Engagements that we did not know about until after we had returned from a mission often occurred between Luftwaffe and Allied fighters. It was really bad news to see German fighter planes with no protective Allied fighters in sight.

Even though we feared fighters more than anti-aircraft fire, more bombers were lost to flak than fighters in 1944 (Sion 2007, 109). The enemy devoted tremendous resources to both the technology and the manufacture of anti-aircraft weaponry. Belts of such defenses were in existence across Germany. The enemy also moved anti-aircraft cannons and guns by rail among potential targets that they wished to protect. Lead planes were a choice target because bombers within a formation often would drop bombs based on the drops of the lead plane. Apparently, the enemy had little problem tracking the speed and path of our formations. Our elevation constituted a different problem for them. Even though their exploding shells were above or below us initially, sooner or later they would learn our elevation.

Anti-aircraft fire was present on every mission I flew. A direct hit could bring down a bomber; an exploding shell could throw off fragments (flak) capable of ripping the fuselage. I observed a direct hit on a bomber on our wing on an early mission. There was a blinding flash, debris floated in the air, but the B-17 was no more. Floating flak typically would not bring a bomber down but would inflict varying degrees of damage. B-17s had considerable capacity to withstand damage; we returned from several missions with three, rather than four, operable engines. I recall counting 87 flak holes in our plane after one mission when flak had been heavy. I knew there would be many flak holes because, while in combat, I could hear pieces of flak striking our plane. In this particular instance, three of four engines were operable when we landed.

HOW GREAT WERE THE RISKS, AND WHAT WERE THE ODDS?

Roger Freeman, in his *History of the U.S. Army Air Force,* wrote that the Eighth Air Force in England faced the worst odds of any U.S. fighting group in any of the services. Bill Wilkins, in a review of Robert Grilley's book, states, "An appendix on American strategic bombing reports that between August 1942 and May 1945, 4,754 B-17s and 2,112 B-24s were lost and that the Germans assigned more than two million men to their anti-aircraft defense while devoting between 30 and 40 percent of their ammunition production to that purpose. The Eighth suffered 46,456 casualties, not counting prisoners." And Sion writes, "Of 350,000 airmen stationed in England with the Eighth, 26,000 were killed, which was the highest percentage of mortality of all the service branches in World War II" (121). But it was Sion's uncle, Richard R. Ayesh, who narrowed the analysis to the odds that faced those who flew and fought. He is quoted by Sion as saying, "Early in the War, only one out of three airmen survived. Then later, with fighter escorts, two of three survived" (119).

Odds improved after D-day, when the Allied forces achieved air superiority. Although Germany improved its technology until it surrendered, it became increasingly difficult for them to compensate for their decline in fighters with improved anti-aircraft defenses. I do not recall discussing survival odds with others who were flying missions when I was doing so. But there were other personnel who, by their behavior and attitude, revealed they were exceedingly conscious of such matters. I refer in particular to the ground crews who maintained our planes. Each crew was assigned a primary plane that would be flown by that crew when schedules meshed. That is, we would fly our assigned plane if it was in condition to fly. It would also be flown by other crews, if we were not flying and their assigned plane was not airworthy. The ground crews who maintained planes knew numerous airmen who had not made it; they also had observed the number of replacement planes required to keep bomb groups in the air. Those ground crews made our planes as safe and comfortable as it was humanly possible to make them. They were available before missions to do their best to meet any request we might make of them. Although I do not know that they did so, it would have been possible for ground crews to calculate survival odds for airmen and planes. I suspect they did so and were motivated accordingly.

I turn now to the unusual history of the bomb group to which we were assigned. The attitude of our ground crews may well have been influenced by that history.

THE "BLOODY 100TH" BOMB GROUP

Shortly after we arrived in England, we learned we were assigned to the 100th Bomb Group. The 100th was known by those familiar with Air Force operations as the "Bloody 100th." I had not heard of the "Bloody 100th" before we were assigned there, but one picks up on such matters fast.

The oral history I first heard pertained to the early history of the Eighth Air Force in England. It involved a 100th Bomb Group B-17 that was returning to England alone from a mission when it was intercepted by Luftwaffe fighters. The bomber lowered its landing gear, a sign of surrender. When German fighter planes came along side, the

B-17 opened fire, and the German planes were shot from the sky. According to this account, the 100th Bomb Group henceforth became a favorite target of the Luftwaffe. Sion adds possible qualifications to this oral history (Sion 207, 65). One is that the intercom equipment on the B-17 may not have been working properly, and the gunnery crew did not know the landing gear had been lowered, signifying surrender. Another is that the B-17 electrical system was damaged and the landing gear lowered without crew awareness.

There is controversy about whether any such incident ever occurred. Luftwaffe survivors were interviewed after the war and denied particular Allied bomb groups were targeted in their attacks. Further, similar stories were told about the 450th B-24 Bomb Group as were told about the 100th Bomb Group. The contention is that such stories served German propaganda purposes. However, a B-17 100th Bomb Group survivor confirms the explanation of the "Bloody 100th" Bomb Group name, according to Michael P. Faley, 100th Bomb Group historian. It makes little or no difference to a description of my journey whether the "bloody" moniker stemmed from myth or reality. The following statement can be found in a history of the 100th Bomb Group Eighth Air Force: "Although the 100th did not have the highest over-all loss rate of any group in the Eighth Air Force, it did have heavy losses during eight missions to Germany, thus earning the nickname 'The Bloody 100th.' " My crew was assigned to the 100th Bomb Group because of its heavy losses on September 11, 1944. And we were on the mission when the 100th again suffered heavy losses on December 31, 1944. The following table provides information about losses on several missions.

Date	Mission	Planes Lost
August 17, 1943	Regensburg–Africa shuttle	9
October 8, 1943	Bremen	7
October 10, 1943	Munster	12
March 6, 1944	Berlin	15
May 24, 1944	Berlin	9
September 11, 1944	Ruhland	14
December 31, 1944	Hamburg	12

My crew, and others assigned to the 100th in September 1944, were 100th Bomb Group replacements for the 14 planes and crews that were lost on the September 11, 1944, Ruhland mission. When we first went to our assigned barracks, a radio was tuned to a German propaganda station broadcasting in English. We were welcomed to the Bloody 100th and promised a fate similar to that of those we were replacing. I was to remember that promise about four months later, when returning from the December 31, 1944, Hamburg mission. The 12 planes lost on the Hamburg mission were the most lost on any mission I flew, but we often returned with one, two, three, or four fewer planes than were with us when we took off. Even so, our odds were far better than those of crews that flew in 1943 and early in 1944. In 1943, a tour of duty required a total of 25 missions, and the odds were very much against completing 25 missions. When we went into battle in September 1944, a tour of duty had been increased to 35. Even at 35, our odds were better than those that faced those crews with a 25-mission requirement. By V-E Day in April 1945, I had flown 30 combat missions.[4] Just before, and shortly after, V-E Day, we flew six food-relief missions in addition to the 30 combat missions. We performed food drops to people in Belgium and the Netherlands (lowland countries), which had been short on food for months as the superpowers were striving for a knock-out blow.

Periods of intense activity often were followed by several days without missions. The first six missions were flown over a relatively short period, from October 15 to November 2. We did not fly again until December 24 because we became a lead crew after our sixth mission. As lead crew, our pilot, in particular, was subjected to special training. At that time our crew was reconstituted. Our officers, except for our pilot, were sent to other crews, as was an enlisted waist gunner. Our remaining crew members were reduced to six, as contrasted to the ten we began training with at Dyersburg, Tennessee. This meant that our lead crew was supplemented with lead crew personnel on a mission-to-mission basis. The breaking up of the crew that we had trained with and with whom we had flown six combat missions, together with a flying pattern of intense activity followed by periods of idleness, was clearly a price we paid for becoming a lead crew.

In the latter part of December 1944, there was an incredible span of good weather. Our bomb group flew almost daily, and my crew flew December 24, 25, and 31. After the difficult December 31, 1944, mission, our principal pilot was relieved as pilot of our crew. We were not told the reason. We then did not fly again until January 6, 1945. We were fortunate to have Lt. (later Capt.) David Raiford assigned as a replacement lead pilot. From that day forward, we flew steadily until the war in Europe ended.

We were proud of being a lead crew. I enjoyed the responsibilities I had as a lead crew radio operator. These tasks relieved stress and monotony on long flights to and from a target. Bomb groups rotated wing leads, and wings took turns leading missions.

All missions were not created equal. There were "milk runs," such as our three missions to Cologne, where we bombed the railroad terminal with little interference except for some flak. In contrast, oil refineries, submarine pens, and aircraft factories were well protected. Nevertheless, some risk existed on every combat mission.

Two missions weigh heavily in my memory. I am not confident of either the date or target for one, although I believe it to be the mission to Nuremberg on February 21, 1945. We had completed the bomb run, and I had sent my coded "bombs away" message for our group. Our crew chief then called me on the intercom and described a difficulty facing us. Our bomb-bay doors had failed to retract after the bombs had been dropped, and they could be closed only by hand cranking. His station was just forward of the bomb bay; mine was just behind. The crank connections for closing the bomb-bay doors were on his side of the bomb bay, but he could only turn the crank with difficulty. To turn it, he had to lean far over the bomb bay to reach the handle at the end of the crank. He requested I crawl through the bomb bay and assist with the cranking. We were flying at approximately 25,000 feet. The temperature in the bomb bay probably was near 40 degrees below zero. The bomb-bay doors were open, and we were in flight. When we were at our regular stations, we had electric heated suits under our flight clothes, functioning much as an electric blanket operates. Of course, if I were to go through the bomb bay, my suit would not be connected to an electrical outlet. At that altitude, one had to be on oxygen constantly lest one lose consciousness and suffer immediate brain damage. A narrow catwalk (probably 8 to 10 inches in width) ran across the bomb bay. Halfway across, a support ran from the catwalk to the roof of the plane. Because the bomb-bay doors were open, the German countryside was visible below. I tried to avoid that sight as I crossed the catwalk.

If I were to help the crew chief, I needed to have a parachute and an oxygen bottle with me. Because my heated suit was not connected, the cold wind quickly penetrated my regular flight clothes. I soon began to shiver uncontrollably. I developed a position on the catwalk from which I could turn the crank. The crew chief and I then alternated turning the crank. It was all either of us could do to move the bomb-bay doors. I would crank as long as I could and be bathed in sweat next to my skin before passing the crank to him. The sub-zero wind would then penetrate my clothing, and I could feel sweat turning to ice. We closed the doors eventually, and I crawled back to my station. I have relived that horrible experience many times. Even today, I shudder as I recall these details.

The other memorable mission occurred on December 31, 1944, when the 100th Bomb Group suffered its last major mission loss in World War II. The official record states that 12 B-17s, or one-third of our group, were destroyed. Until I saw the official record, I thought the loss was even worse because I could count only one-half of our planes in formation after the fighter attack. The difference can be explained by bombers being separated from our group during the attack, then entering another group's formation for the return to England. Others may have found their way home independently, although that is unlikely because of the vulnerability of individual planes. Our fighter protection was delayed getting off the ground in England because of weather. We were hit by German fighters after we had finished dropping bombs on oil refineries and submarine pens at Hamburg. Our tail gunner was given credit for shooting down a fighter plane, and our top turret gunner was given partial credit for one during the December 31, 1944, battle. The B-17 radio operator had a single 50-caliber machine gun in the top of the plane, with limited visibility. One German fighter plane came into my view. I discharged a few rounds in its direction. There was no reason to believe I did damage. Two members of our original crew were on a plane that went down that day. I knew the location of their plane and saw two parachutes open as the plane fell from the sky. I learned subsequently they became prisoners of war and were released at the war's end. Not everyone was so fortunate; many perished.

As I reflected on that day, and later read accounts of what transpired, I came to recognize why it is so difficult to obtain an accurate assessment of battles in war. Much often happens in a short time and in different locations. Most participants are intensely focused on events in which they are involved, and they may have little knowledge or distorted impressions of other events. For example, an article was recently called to my attention that described a midair collision of two B-17s during the fighter attack on the December 31 Hamburg mission (Flatley 1997). The two planes came together, one above the other, and flew joined until they went to the ground in Germany. Except for the pilot and copilot of the upper plane and a ball turret gunner who perished, the crew members bailed out. When the two planes that were joined went to the ground, the lower plane exploded, and the other plane broke away. The pilot and copilot survived, were taken prisoner, and lived to discuss the incident. Although I was a participant in that battle, I knew nothing of that remarkable event until I read the article referred to above.

The December 31, 1944, mission was a surreal experience. In retrospect, I have difficulty bringing my recollections into focus. I have thought a great deal about those who flew B-17 missions in 1943 and early in 1944. The survivors would have had more than one experience, in fact several, similar to that of December 31, 1944. When I think of those who flew during the earlier period, I feel both gratitude and guilt: gratitude for what they did; guilt because I was so much more fortunate than they.

REFLECTIONS

Reflections about my combat experience fall into two categories. One set pertains to personal reactions. The other category involves public policy implications of strategic bombing.

MY REACTIONS: As I thought about combat prior to entering the Armed Forces, I wondered if I could bring myself to inflict damage on others, whether because of compassion or because I would be so frightened I could not function. The reality was different from either of my conjectures.

The air warfare I waged was more impersonal than I had imagined. We did not see our victims. The fighter planes we tried to bring down were trying to bring us down first, and, anyway, they flew by at enormous speed. I felt little compassion dropping bombs on factories, refineries, submarines pens, or railroad terminals even though, on reflection, I knew people probably were being killed in the process. I was troubled by, and continue to feel remorse about, bombing that was carried out to damage morale by destroying residential areas and civilians in those places.

Did I experience fear? You bet, but I was never so frightened that I was unable to discharge my responsibilities. The Army Air Force trained me well. I knew the roles I was supposed to play and how to play them. Clearly, I was fit physically when flying both training and combat missions. I never vomited or had digestive problems. I had difficulty sleeping, as I recall, only after the Hamburg mission and the catwalk bomb-bay experience. I had dreams about the Hamburg mission for a few years. The bomb-bay event still returns in vivid detail from time to time.

Yet there have been lasting psychological effects. I believe war to be immoral and dehumanizing. I do not go to war memorials and military parades if they can be avoided. I have never joined a veteran's organization. And I am told I talk little about combat events. Do I believe war is ever justified? Perhaps, although I am convinced violence is used to resolve differences far more than is justified. I do not believe the incentives affecting decision makers in the making of war and peace are appropriate even in democracies. Wars make fortunes and create enduring legacies for politicians, military leaders, and others. The alternatives to war do not confer comparable rewards.[5]

Because of a lack of emotional and intellectual development, I did not take full advantage of the three years of my life that were dominated by the armed services. This was especially true of the time I spent in England, when I had considerable free time. I did enjoy greatly the English countryside, the villages, and the cities of London and Edinburgh. From my standpoint, the English people were wonderful, and I learned something about their culture firsthand. Even so, there were many things I should have done that I did not do. Much of the fault was mine, but the Army Air Force could have done more. Library facilities on my base were nearly nonexistent. Little information was provided regarding what might be available in the community or, say, in London when we were there. Most of us were 19 to 22 years of age, socially immature, and needed to learn about alcohol, sex, gambling, and society generally. It seemed to be assumed that if we could manage the Eighth Air Force over Germany on complicated missions, we surely must know about the birds and the bees, flora and fauna, geography and history. Neither at that time, nor in retrospect, did I believe officers were much ahead of enlisted men in such matters.[6]

STRATEGIC BOMBING AND PUBLIC POLICY: Air power developed greatly during World War II. The United States, Great Britain, the Soviet Union, Germany, and Japan, in particular, invested heavily in aircraft and personnel to wage war in the air. An obvious and early use of air power was for tactical purposes, in support of existing ground and naval operations. It then became apparent that aircraft could be made to fly long distances and carry heavy loads, and with these developments, strategic use of air power became a possibility. Strategic bombing during World War II was controversial and became a public policy issue. The following quotations are included here to provide an appreciation of this issue.

> Strategic bombing is best defined as the use of air power to strike at the very foundation of an enemy's war effort—the production of war material, the economy as a whole, or the morale of the civilian population—rather than as a direct attack on the enemy's army or navy. A strategic air campaign almost always requires the defeat of the enemy's air force, but not as an end in itself. While tactical air power uses aircraft to aid the advance of forces on the ground or on the surface of the ocean, usually in cooperation with those forces, strategic air power usually works in relative independence of armies and navies, although its effects may complement those of a naval blockade—the

operation of war most comparable to strategic bombing in its attack on the sources of enemy power (Levine 1992).

The following quotation is from the writings of John Kenneth Galbraith:

On three nights in late July and at the beginning of August 1943, the heavy planes of the RAF Bomber Command droned in from the North Sea and subjected the city of Hamburg to an ordeal such as Germans had not experienced since the Thirty Years' War. A third of the city was reduced to a wasteland. At least 60,000 and perhaps as many as 100,000 people were killed—about as many as at Hiroshima. A large number of these were lost one night when a ghastly "fire storm," which literally burned the asphalt pavements, swept a part of the city and swept everything unto itself. Adolph Hitler heard the details of the attack and for the only known time during the war said it might be necessary to sue for peace. Hermann Goering visited the city with a retinue to survey the damage and was accorded so disconcerting a reception that he deemed it discreet to retire.

Yet this terrible event taught a lesson about the economics of war which very few have learned and some, indeed, may have found it convenient to ignore. The industrial plants of Hamburg were around the edge of the city or, as in the case of the submarine pens, on the harbor. They were not greatly damaged by the raids; these struck at the center of the city and the working class residential and suburbs. In the days immediately following the raids production faltered; in the first weeks it was down by as much as 20 or 25 percent. But thereafter it returned to normal.... the efficiency of the worker as a worker was unimpaired by this loss. After a slight period of readjustment, he labored as diligently and as skillfully as before (Galbraith 1958).

When I was flying bombing missions, it never occurred to me there would be controversy as to whether we were advancing the Allied war effort. I was aware that air power had been questioned in the armed services between World Wars I and II, as reflected in the court-martial of Billy Mitchell. Yet I assumed air superiority was recognized as important, and that strategic bombing was considered to be part and parcel of air power generally. It was not long after World War II ended, and I had become an economist, that I was told by another economist about the United States Strategic Bombing Survey. I was told it had been prepared by several economists under the chairmanship of John Kenneth Galbraith. Further, I was told they concluded that our hard work in the Eighth Air Force bombing Germany had not paid off. It is not surprising that someone with my background was skeptical of any such generalization. As it turned out, my skepticism was justified. What I had been told was an oversimplification of the survey findings. And, with the passage of time, a better-balanced appraisal could be made (Grant 2008).

I refer to this controversy because it raises issues that I have encountered repeatedly in my intellectual journey. Some of the substantive findings in the survey about the strategic bombing are of little relevance currently because both economies and armed conflicts have changed greatly. Yet methodological issues that were raised are of relevance in both war and peace. Some of the survey findings were not surprising. There is now a consensus that control of the air by the Allied forces by the time of D-Day was of great consequence. There may be disagreement as to the precise contribution of strategic bombing in bringing this about. After D-Day, strategic bombing clearly damaged oil refineries, aircraft production, and transportation. The survey results, however, made clear that morale of the Germans was not greatly affected because there was a cushion of certain goods that the bombing did not affect. Further, there was excess capacity in the German economy early in the war, and bombing did not cause that to disappear.

It was after D-Day that strategic bombing reached its height. By that time, the Allied forces had air superiority. Both the survey and its critics agree that during this period, bombing did cause harm to oil refineries and submarine production and operation, as well as to transportation. Furthermore, German resources—human power, ammunition, and equipment—were employed in the protection of key targets from strategic bombing. There was less agreement regarding damage to morale of civilian populations from strategic bombing at that time. The Levine quotation at the beginning of this section states that one objective of strategic bombing may be to strike at the enemy's war effort by damaging the morale of the civilian population. This might be accomplished by destroying neighborhoods, by reducing consumer goods production, or by direct bombing of civilian populations. The nuclear bombs dropped on Hiroshima and Nagasaki are examples of the latter strategy. And these examples help us understand why it is that strategic bombing has remained a controversial issue in every armed conflict in which the United States has engaged since World War II.

In the Levine description of strategic bombing, the sole objective, although implicit, is to defeat the enemy. Such an objective may be the sole objective for some, but not necessarily for everyone. Those for whom defeat of the enemy is the sole objective will be concerned with direct bombing of civilians only to the extent that their morale is damaged and the war effort diminished in some way. The Galbraith quote draws direct attention to the ineffectiveness of strategic bombing in affecting civilian morale. Defeat of the enemy may be one, but not necessarily the sole, objective of others. Assume that they wish to avoid the loss of civilian lives, but, nevertheless, wish to win a war. In that case, they would wish to use strategic bombing differently than it was used by the RAF at Hamburg and as described by Galbraith. Particular evaluation standards are required for each distinct set of strategic bombing objectives. Possible policy objectives would include, but would not be limited to: 1) abandoning strategic bombing altogether on the grounds that there are more effective ways to win wars, 2) embracing strategic bombing with no constraints, and 3) embracing strategic bombing but not using it as an attempt to damage civilian morale.

As supreme commander in Europe, General Dwight Eisenhower made a decision that assigned night bombing to the Royal Air Force (RAF) and daylight strategic bombing to the U.S. Eighth Air Force. It can be said that, for the most part, both were engaged in strategic bombing. The difference was that the RAF emphasized bombing population centers in an attempt to damage civilian morale. The Eighth Air Force favored transportation facilities, submarine pens, oil refineries, and other industrial works as targets. Both the United States and the

United Kingdom wished to win the war. But the civilian population of the United Kingdom had suffered from blitzes and bombs fired by the Germans from the Continent. The two Allied nations were not of one mind in their desire to deliver "payback" by damaging civilian populations. Further, neither nation had accurate a priori expectations of the effects of strategic bombing in winning wars.

This consideration of strategic bombing brings a fundamental characteristic of many public-policy controversies into the open: *One party's ends may be another's means.* To the extent this remains true, expert analysis will not cause controversies to disappear. It has been said that nothing gets settled in Washington, D.C. Rather, it is a place where certain issues are continuously revisited. This is not to say that expert analysis is of no value in public policy formation. It can change the nature of debates by illuminating certain issues, but standing alone it will not resolve basic controversies. I return to this conundrum later in this book.

3

AT LOOSE ENDS
(1946–1954)

FROM AIRMAN TO CIVILIAN

On May 8, 1945, the enemy in Europe surrendered unconditionally. Much has been written about how rapidly the United States Armed Forces had been organized, equipped, trained, and deployed into combat. Yet the rapidity with which redeployment occurred after V-E Day and V-J Day was equally, if not more, impressive. The processes that affected me were occurring simultaneously with others in all branches of the armed services.

Approximately three weeks after V-E Day, my crew was assigned to ferry a B-17 back to the United States. Our route was almost the same that we had taken in the opposite direction in 1944. One difference was that we landed in Iceland for a few hours of sleep on the return trip. I was assigned to a troop train in New England that returned me to Fort Leavenworth, Kansas. I was then given a 30-day furlough before being sent to the West Coast. My military occupational specialty, aircraft radio operator, was in short supply. I was told I was ticketed to go to the Pacific Theater, where I would be trained for flying on aircraft used there, probably B-29s. And then V-J Day, August 15, 1945! My assignment was changed, and I was sent to an air base near Liberal, Kansas, where I received my honorable discharge November 2, 1945. In less than six months, I had helped ferry a B-17 across the North Atlantic, taken a 30-day furlough, gone to the West Coast where I expected to be sent to the Pacific, but then my world changed and I was discharged! Within the three months that followed, I would be married and make an occupational decision that affected my intellectual journey for the remainder of my life.

During World War II, my parents had used their modest lifetime savings to purchase a farm in Montgomery County, Kansas, which became my new home. This was a small, diversified farm where they lived until my father's death about five years later. They did not have adequate capital to make it a highly profitable operation, but for the first time in their farming experience, they were owners and did not have to satisfy a landlord. At that time, my younger brother, Don, was a senior in the Independence, Kansas, public high school. He was a good student and, after graduation, attended Kansas State, where he obtained a degree in chemical engineering.

While I was flying combat missions, I had written to my former high-school teacher, Merab Weber, to ask if she would correspond with me. Merab left teaching during World War II for employment at Boeing Aircraft in Wichita, Kansas. She agreed to correspond, but after I was discharged, I went to see her. We became attracted to one another and were married January 20, 1946. As I was trying to decide how I was going to make a living as a civilian, Merab had no apparent reservations about casting her lot with a confused, insecure guy. She never complained about living in small, poorly heated or ventilated places as I was acquiring an education. And, throughout our marriage, she made all the necessary adjustments to benefit me professionally. When we went to Manhattan, Kansas, to live after our wedding, we found housing to be limited because of the many returning servicemen who were seeking an education. We were able to obtain a two-room upstairs apartment that was miserably hot during Kansas summers. Our kitchen consisted of a sink and cook-stove crammed into a bedroom closet.

As soon as Merab and I decided to be married, I faced the same problem as many other discharged servicemen across the country: How was I going to support myself and my wife? I explored employment possibilities, but advancement opportunities appeared limited or the work seemed uninteresting. I then became aware of the GI Bill, which provided subsidized education for veterans. I investigated attending the University of Kansas (KU) and Kansas State (K State). I intended to study business and law if I attended KU. I went there to explore possibilities and received the impression that administrators at KU were not excited about accommodating one more returning GI. Administrators at K State had my records and encouraged me to return there. I was not clear what I might do upon graduation with a degree in agriculture, but I was assured there would be employment opportunities. K State recognized the academic credit for the courses I had taken earlier, both in correspondence and in residence. I enrolled at K State in January 1946, shortly after our wedding on January 20. In retrospect, I recognize there were two important influences that interacted to cause me to take this action. One was the attitude of my family, especially Dad, regarding education. The other was the GI Bill. It turned out I was to milk the GI Bill for support for three academic degrees. I shall be forever grateful for the opportunities that piece of legislation has afforded me.

THE RETURN TO K STATE

I was in a hurry to begin my life's work, although I did not know what that could or would be. I did not return to K State motivated by a great desire for learning and knowledge. I knew acceptable grades would be helpful when the time came for me to obtain employment. I decided a B average would achieve that objective. I then crowded as many courses into my schedule as I thought consistent with that objective.

A personal intellectual awakening occurred in the summer of 1946. I enrolled in a course named Applied Economics taught by Professor Edgar Bagley. He developed one or more analytic models for several public policy issues and then used those models to isolate questions for class discussion. Monetary and fiscal policy, labor economics, international trade, and agricultural policy are examples of policy issues we discussed. Mastering those models and then discovering their relevance to public policy "hooked" me from that time to the present. The final examination for the course turned out to be an ego-boosting experience. It was a two-hour examination consisting entirely of multiple-choice questions. I enjoyed the examination and believed I had

Merab Weber Castle; Emery Castle. Photographs taken around the time of our marriage.

performed credibly. Ed Bagley saw me in the building where his office was located a few days later. He asked me to come into his office, and then asked where I had taken Principles of Economics, a prerequisite for his Applied Economics course. I told him I had taken it as a correspondence course prior to World War II. He said, "That explains it. I have given that examination to many students. You obtained the second best score anyone has received. You answered only two questions incorrectly, and they were the two most elementary questions on the test. No one has ever obtained a perfect score on that test, but you and the person who missed only one question a few years ago came close." It was then that I first considered becoming an economist.[1] It was my good fortune that I was able to take additional courses from Bagley while completing the remainder of my undergraduate degree work and when I was working for my masters.

Another remarkable intellectual experience occurred during my undergraduate work at K State. Milton Eisenhower, president of K State, had previously introduced comprehensive courses in that technically oriented educational institution. Comprehensive courses covered subject matter not provided for in technical, specialized curricula. These comprehensive courses included Biology in Relation to Man, Man and

Cheryl Diana Castle, age 2   Cheryl Diana Castle, high school graduate

the Social World, Man in the Physical World, and Man in the Cultural World. These courses typically covered subjects from a multiple-discipline perspective and in a historical context.

I took Man and the Cultural World during my senior year. The reasons I did not take the other comprehensive courses were complex. In the first place, I was not expected to take Man in the Social World because I was an economics major. I was in an agricultural curriculum and for that reason did not take the biology comprehensive course because I took, or would take, several courses in biology. I did not take the physical science comprehensive because of scheduling difficulties. I am indeed grateful it was possible for me to take Man and the Cultural World.

In that course, we considered major contributions of the arts, sciences, and humanities during particular historical periods. The lectures were team-taught. A philosopher, a musician, and a novelist, for example, would describe, or reenact, how intellectual contributions made during a period were interrelated. Students, with help from professors in recitation sessions, discussed connections across fields and disciplines. My recitation instructor was a historian and a marvelous integrator. This course opened my eyes to a world I knew almost nothing about. These courses were not well accepted at K State and were discontinued immediately after Milton Eisenhower left to become president of Pennsylvania State University. He subsequently became president of Johns Hopkins University. Based on conversations I have had with people who were then faculty at Johns Hopkins, I conclude that it was a marriage made in heaven. In any case, I am exceedingly grateful for the one truly "liberal arts" course I had at K State. It has served me well for more than six decades.

I completed my work for the baccalaureate in January 1948, having been in college continuously since reentering K State two years earlier. I did better than my B average objective. In fact, I graduated with honors, something I did not know about until I saw the commencement program. I was the first person in my family to obtain a degree from a four-year college. My parents came to the January commencement and, needless to say, were proud and delighted. They were pleased to learn they would be grandparents, if all went well, the next summer. Merab had done substitute teaching and clerical work for nearly two years, prior to becoming pregnant.

When my parents were in Manhattan, I told them my department had offered me a graduate assistantship if I wished to work for a masters degree. The stipend I would earn, together with the GI Bill support, would increase our monthly income. My parents, especially Dad, believed education for poor families was a great thing; they supported the decision Merab and I made to continue my education. I began to work for my masters immediately after obtaining my bachelors.

A FLEDGLING ECONOMIST

I began taking courses from Bagley again. This time it was economic theory. We read Alfred Marshall, Edward Chamberlin, Joan Robinson, and J. M. Keynes. The department faculty believed that economic theory was important, but there was little direct connection between economic theory and the research program of the department. This soon became a major issue in agricultural economics largely because of the scholarly work of Theodore Schultz at the University of Chicago and Earl Heady at Iowa State. Both of these people later became mentors of mine, but at that time they were not aware of my existence. I used a model from Marshall's principles to guide empirical work in my masters thesis. This attracted considerable attention in my department. Later, I published my first journal article, a note in *Land Economics* based on my thesis (Castle 1952).

I taught a recitation section of the Principles of Economics course as part of my assistantship responsibilities. Institutions of higher education were then still struggling to accommodate the large number of veterans attending college aided by the GI Bill. I had mostly engineering students in my recitation section. At our first meeting, I told them I had no teaching experience, and there was much about economics I did not know. I also told them that if I did not know something during class, I would tell them so, and provide an answer at our next meeting. They responded most positively. I learned there is no disgrace in saying "I don't know" and concluded that both science and economics would benefit if this were practiced more widely. Although I was not well prepared for this teaching responsibility, I was greatly stimulated by it. I began to think that helping others learn about economics might become a life's work and a way to provide for my family.[2]

To my surprise, I was promoted to a full-time instructor when I completed work for the masters degree. The K State Department of Economics and Sociology had dual responsibilities. It reported to the School of Arts and Sciences for instruction in economics and sociology and the School of Agriculture for research and instruction in agricultural economics. I was delighted by the promotion. I would be paid to do two things I really liked—learn economics and teach. But there was a catch—I came to learn that a doctorate would be necessary if I were to advance in the academic world. This had not always been the case at K State, but the word came down that beginning faculty would be expected to obtain the doctorate.

Merab and I discussed this problem at length and decided that if I intended to teach, it would be better to work for the doctorate sooner rather than later. I did not decide to pursue the doctorate because I thought of myself as an outstanding scholar or because I wanted to do research. My motivation was to qualify for a faculty position at K State or a similar institution. When I told Dad of our decision, he was supportive, but he may have thought I was overdoing the education thing a bit.

Edgar Bagley`   Earl Heady

IOWA STATE UNIVERSITY

Earl Heady, a young faculty member at Iowa State University, was acquiring superstar status in agricultural economics when I decided to go there in 1950. He used the theory of the firm, as set forth in neoclassical economic theory, and statistical analysis to revolutionize teaching and research in farm management and production economics. He was attracting outstanding graduate students from throughout the United States and abroad. I decided to go to Iowa State after considering the Universities of Wisconsin and Chicago. The graduate program in economics at Iowa State was rich in tradition. Theodore Schultz, who was to become a Nobel Laureate in economics, had earlier provided outstanding leadership before moving to the University of Chicago. Schultz had attracted to the Iowa State faculty such outstanding economists as George Stigler, Albert Hart, Kenneth Boulding, and Gerhard Tintner. Tintner was at Iowa State when I did graduate work there.[3] I relied heavily on Hart's writings in my doctoral dissertation research, and I came to know Boulding in later years. Stigler also became a Nobel Laureate in economics after he left Iowa State. He went to the University of Chicago, where he joined his good friend Milton Friedman in the economics program.

Iowa State was similar to K State in that both economists and agricultural economists were in the same department. Thus, throughout my formal education, my primary department was the academic home for both economists and agricultural economists. I thought of myself as an economist specializing in agricultural economics, a view that has persisted throughout my career.

My economics course work at Iowa State was largely rigorous neoclassical economics. It emphasized J. R. Hicks and ordinal utility, rather than Alfred Marshall and cardinal utility, as was the case at K State. Cardinal utility may be thought of as an indicator of levels of satisfaction associated with various quantities of goods and services, originally referred to as "utils." Marshall made no attempt to place precise measures on this indicator and did not believe it was possible to make interpersonal utility comparisons. J. R. Hicks was among the first to develop relative preferences that might be assigned to varying quantities or combinations of goods and services, an approach commonly referred to as "ordinal utility." The "new" welfare economics was developing, and we were expected to master this literature. Although I thought I might be "in over my head," I learned a great deal and, more importantly, realized there was much I did not know.

The greatest weakness of my graduate work in economics was that it did not provide opportunity to learn how economic doctrine had developed. In retrospect, I recognize that if I had known more about the early development of economic doctrine, I would have had a more informed opinion about what I did and did not know. My education to that time had been in the mainline Western economics tradition descending directly from Adam Smith and Albert Marshall. In the following section, I present a summary description of that tradition in economic science.

THE ENGLISH-AMERICAN TRADITION IN ECONOMIC SCIENCE[4]

We begin with Adam Smith. Yet how does one provide a "summary description" either of Adam Smith as a philosopher or his thousand-page opus, *An Inquiry into the Nature and Causes of the Wealth of Nations*? Think also about the contributions of numerous other economists since 1776. The only correct answer is: "very superficially."

Consider first Figure 1.[5] The economic thought discussed here stems from the works of Scottish-born Adam Smith. All histories of

FIGURE 1

ECONOMIC THOUGHT FROM ADAM SMITH TO THE PRESENT

ADAM SMITH
(1723–1790)

◠ ECONOMIES ◠		EUROPEAN ECONOMISTS	NONESTABLISHMENT AMERICAN ECONOMISTS
EQUILIBRATE	**EVOLVE**		
CLASSICAL			
David Ricardo (1772–1823)	Thomas Malthus (1766–1834)	François Quesnay (1694–1774)	
John Stuart Mill (1806–1873)		AntoineAugustinCournot (1801–1877)	
William Stanley Jevons* (1835–1882)		Karl Marx (1818–1883)	
NEOCLASSICAL			
Alfred Marshall (1842–1924)		Leon Walras* (1834–1910)	Henry George (1839–1897)
John Maynard Keynes** (1883–1946)		Carl Menger* (1840–1921)	Thorstein Veblen (1857–1929)
Frank Knight (1885–1962)	Allyn Young (1876–1929)		
Edward Chamberlin (1899–1967)	Nicholas Kaldor (1908–2006)	Vilfredo Pareto (1848–1923)	
Theodore W. Schultz (1902–1998)	John Kenneth Galbraith (1908–2006)	Johan Wicksell (1851–1926)	
J. R. Hicks (1904–1989)	Kenneth Boulding (1910–1993)		
Milton Friedman (1912–2006)			
Paul Samuelson*** (1915–)			
Kenneth Arrow (1921–)			
Robert Solow (1924–)			
NEW GROWTH THEORY Paul Romer (1953–) Paul Krugman (1955–)		* Marginalist pioneer ** Macro revolutionary *** Mathematical revolutionary	

economic thought need not have this orientation, but that is the one I experienced, and therefore it is appropriate here. The figure is based on the proposition that there were two important perspectives adopted by Adam Smith in the *Wealth of Nations* (Warsh 2006): One was how economies *evolve* with the passage of time. The other was that individuals, firms, industries, and entire economies *equilibrate* among various forces in the process of wealth creation. Given the complex nature of these perspectives, it is not surprising that economists who followed Adam Smith came to be identified with one or the other perspective. Such classifications are necessarily arbitrary and approximate. These were and are people of great intellectual capacity, and often they were concerned with both *evolutionary* and *equilibrating* forces.

Remember also that Adam Smith did not create the *Wealth of Nations* in a vacuum. He was in his fifties when it was published, and it was preceded by his *Theory of Moral Sentiments* in 1759. Smith traveled in Europe and came to know personally many economists there. Nations had long engaged in international trade, and Smith had studied the consequences of such trade. He had observed economic processes in the countryside, as well as in towns with small businesses and cottage industries. He had been a student at Oxford, although critical of the education he received there. He was familiar with the philosophical treatises of the time and knew personally several contemporary philosophers. He came to believe that one of the deficiencies of the philosophy he read was the prevailing negative view of self-interest. He had concluded that there were both personal and social benefits that stemmed from actions individuals took to advance their own interests. He captured this belief with his coinage of the term "the invisible hand," one of the most enduring in the English language. It is less well known that he used the term only once (deep in the body of the *Wealth of Nations*) and qualified extensively the use he made of it. Because his coinage has become so well known, it is often assumed that he was typically critical of government intervention in economic affairs. Yet Jacob Viner, a noted scholar of economic doctrine, concluded, "Adam Smith was not a doctrinaire advocate of laissez-faire. He saw a wide and elastic range of activity for government."

As noted, Adam Smith was familiar with the writings of the European economists and, in fact, knew several personally. Smith said he would have dedicated the *Wealth of Nations* to the French economist François Quesnay had Quesnay been living in 1776. In addition to Quesnay, many European economists of various nationalities influenced many English-American economists. Some are noted in Figure 1, column 3.

THE CLASSICAL ECONOMISTS: The *Wealth of Nations* qualifies as a revolutionary scientific accomplishment because it attracted subsequent scholars to its elaboration and refinement, who came to be called classical economists. Only four are listed in Figure 1. I call attention initially to Thomas Malthus, David Ricardo, and John Stuart Mill, each of whom elaborated the insights of Smith in a distinctive way. All were born and lived in Great Britain.

Malthus is best known for his essays on population and the means of subsistence, but of greater relevance here is his identification of other subjects that have been of continuing concern in economics. He believed that economic events have more than one cause, but that there is great temptation in economics to ascribe more influence to a single cause than is warranted. He questioned whether an automatic equilibrating device existed in economics, and he believed that gluts as well as scarcities may be created. His views anticipated subsequent economic thought by Karl Marx and John Maynard Keynes.

David Ricardo was a businessman early in life and then was elected to Parliament. He believed economics could become a science underlying the formation of public policy (political economy). He pioneered the use of hypothetical-deductive models in economic reasoning. This permitted self-evident propositions to be used for the construction of models from which deductions could be made about possible future economic events. He often ascribed great influence to a single cause. He was a student of the major factors of production—land, labor, and capital—and he was a major contributor to rent theory. He believed distribution was the real objective of political economy and once wrote, "The natural price of labor is the minimum cost of producing men" (Gray 1931, 183).

John Stuart Mill was born early in the 19th century and was educated largely by his father in both philosophy and economics. He understood well the methodological implications of the hypothetical-deductive model advocated by Ricardo. He believed political economy was concerned mainly with production, but that distribution was the province of society in total and was not governed by iron laws of production. He came to believe economics was an inexact statement of tendencies, until a correct allowance is made for other causes. Mill

continues to be cited in philosophy of science and research methodological treatises in economics. I have learned a great deal from each
of these classical economists and find myself returning to John Stuart
Mill repeatedly.

The other classical economist listed in Figure 1 is William Stanley
Jevons. His contribution is considered here simultaneously with that
of European economists Leon Walras and Carl Menger. These three
economists were contemporaries who independently incorporated the
differential calculus in theories of economic systems. Each of the three
was interested in different facets of economics, suggesting a unifying
theory for all of economics grounded in deductive logic. Walras was
interested in general equilibrium, Jevons wished to express the maximization of pleasure, and Menger worked to express subjective value
at the margin when analyzing demand. Each of these three economists
discovered the power of marginal analysis in economics at about the
same time. Alfred Marshall was also a contemporary who employed
marginal analysis to bring both demand and supply into a common
framework. I studied in the English-American economics tradition
with emphasis on Alfred Marshall.

NEOCLASSICAL ECONOMICS: The origin of neoclassical economics is
often dated at 1890, which was the year of the first printing of Alfred
Marshall's *Principles of Economics.* In this book, marginal analysis is
used to provide a beautiful synthesis of demand and supply from the
perspective of individuals, firms, industries, and an entire economy.
Marshall distinguished between short- and long-run analysis and
noted the symmetry of rent in production and consumers' surplus in
consumption. He placed more emphasis on partial than general equilibrium and disliked long chains of reasoning about economic affairs.
He believed economics should become more like biology than physics, yet his methodology was more akin to physics than biology. He
viewed a competitive economy as the general case and treated monopoly, unequal knowledge, and increasing returns as aberrations. They
were considered as market failures, justifying intervention or special
theoretical assumptions. Alfred's wife, Mary, was also an economist
and supported his efforts throughout her professional career. I read

and studied Marshall's *Principles of Economics* when working for my masters. It has had great influence on my thinking as an economist.

The greatest challenge to the dominance of Marshall did not occur until 1936, with the publication of John Maynard Keynes's *The General Theory of Employment, Interest and Money.* As noted elsewhere in this book, Thomas Kuhn has described a revolution in science as occurring when other scientists are attracted to the elaboration of the discovery in various ways. Keynes's book certainly meets this qualification. As a result, macroeconomics has achieved near-equal standing with microeconomics. Macroeconomics provides a "top down" view of an economy and is concerned mainly with the management of economy-wide aggregates such as national production, price levels, and employment. Microeconomics proceeds from the bottom up, beginning with the decisions of individual consumers and producers.

To this point only passing reference has been made to great economists outside the English-American tradition. This summary reflects the tradition in which I studied to a considerable extent. Admittedly, there are important economists about whom I am not competent to offer comment—for example, Karl Marx as an economic theorist. My limited reading of Marx has made clear to me that he was familiar with, and made use of, classical economic concepts.

All of the English-American economists reviewed to this point did their professional work in England. Yet the American influence became significant late in the 19th century, both within the mainstream of English-American economics as well as for nonestablishment economists outside the mainstream. I comment briefly here about Henry George and Thorstein Veblen. Neither had significant impact on my intellectual journey, yet an appreciation of both contributes to an understanding of the English-American tradition in economics.

Henry George was a self-educated journalist who became a political activist and reformer. His widely read *Progress and Poverty,* published in 1879, was his principal publication. It is not possible to do justice to his wide-ranging view of economics and public policy in a few words. He attributed many social problems to the concentrated ownership of land, and his reforms typically were directed to the redistribution of rent to land. He anticipated the development of environmentalism and developed such concepts as "Spaceship Earth," common property, and the rights of the unborn.

Thorstein Veblen published late in the 19th century and early in the 20th. His work is difficult to classify, although he is frequently referred to as an institutional economist because of his interest in institutional evolution. In his view, institutions were customs, canons of conduct, and principles of right and propriety. He had little use for neoclassical economics as advanced by Alfred Marshall. Some of his most withering criticism was directed at the so-called "economic man." Veblen marginalized himself from the mainstream economic profession for at least two reasons: his critical view of neoclassical economics and his refusal to work within such a system of thought. Despite such views, he held appointments in highly regarded departments of economics at the University of Chicago and Stanford and edited the *Journal of Political Economy* for a period. He paid lip service to both theory and empirical information, but he failed to integrate the two in his own scholarly work, although he made trenchant social observations in his articles and books. Apparently, it was not only reactions to his scholarly work that marginalized him academically; his unorthodox personal behavior played a role in this respect as well.

Some of the most interesting developments in neoclassical economics occurred after the last printing of Marshall's *Principles* in 1920 and before Keynes's *General Theory* in 1936. Allyn Young was one of the most influential economists contributing during this period. He was well known in both England and the United States. He is remembered today largely for his article "Increasing Returns and Economic Progress," which addressed the fundamental determinants of economic progress as set forth by Adam Smith, with special attention given to size of market and decreasing costs. Young supervised two doctoral dissertations that became better known than anything he published and are now regarded as classics. One was Frank Knight's *Risk, Uncertainty and Profit*; the other was Edward Chamberlin's *The Theory of Monopolistic Competition*. He also was the major professor for Nicholas Kaldor, an important economics scholar for much of the 20th century. Although these contributions came to be overshadowed by the work of Keynes, they are of an enduring nature and continue to be applied. I studied both Knight and Chamberlin when I was doing my graduate work and drew heavily on Knight in my dissertation research.

In 1938, Joseph Schumpeter's *Capitalism, Socialism and Democracy* was published. Schumpeter called attention to "creative destruction" processes that occur within capitalism. Schumpeter's significant contributions on the economics of growth and change were also overshadowed by Keynes's *General Theory*, which emerged at about the same time. The Swedish economist Knut Wicksell contributed to some of the same topics, as did Keynes and Schumpeter.

Both Theodore Schultz and J. R Hicks (see Figure 1, column 1) had significant, but very different, influence on my intellectual journey. I came to know Schultz personally after I completed my graduate work but while I was new in the profession. He became interested in my research and writing and served as a mentor until near the end of his life at age 96. I read most of his writings and sought his opinion on many decisions I considered important. He played the same role for many others junior to him. I never met J. R. Hicks but studied his *Value and Capital* when I was doing graduate work at Iowa State. He was a pioneer in advancing the use of ordinal utility, which facilitated creation of the "new welfare economics" that drew on the work of Italian economist Vilfredo Pareto. These developments subsequently became important in resource and environmental economics, a subject I later helped bring to prominence.

I place John Kenneth Galbraith together with Milton Friedman despite their intellectual differences. They were born within four years of one another, and they died in the same year. One came to prominence while a faculty member at Harvard, the other as a member of the University of Chicago faculty. Both became famous as advocates for public policies that were derived from the economic analyses each conducted. Here, the similarities end. The public policies each advocated were very different. Friedman's policy recommendations depended heavily on the accomplishments and potential of decentralized markets. Galbraith was concerned with countervailing power among large organizations: corporations, labor unions, government. Both were excellent writers and appealed directly to the public, as well as to other economists, when advancing policy positions. It is not surprising that Galbraith advised John Kennedy and Friedman advised Ronald Reagan.

I turn next to another revolution in economics of an altogether different nature than the Keynesian revolution. This one was spearheaded by American economist Paul Samuelson and was based on a mathematical rewriting of both micro- and macroeconomics. This began a trend that has transformed the entire economics discipline. Not all of the nuances in economic theory prior to Samuelson were equally amenable to mathematical treatment. Some important concepts were temporarily neglected in the mathematical rewriting, with emphasis placed on those subjects especially amenable to mathematical analysis. General equilibrium analysis provides an example. The rapid growth of mathematical economics coincided with greatly improved capacity for numerical manipulation and processing. This created opportunities for economics to become more empirically oriented, an opportunity many economists eagerly exploited. The mathematical revolution in economics meets all of the requirements of a scientific revolution as set forth by Thomas Kuhn and described elsewhere in this book (see Appendix C).

Next, Kenneth Boulding. I have placed him in the "evolve" column, although his publications, taken as a whole, defy stereotyping. He shared Marshall's belief that economics and biology have much in common and expressed views about the importance of economics becoming more evolutionary. He also was an excellent writer. For many years, I tried to have something he had written nearby so that I could read for either learning or pleasure or both. The learning was not always in economics, but always worthwhile. Ken Boulding often used humor as a sword, a characteristic that did not endear him to some in the profession who took themselves and their discipline very seriously.[6] I came to know Boulding well during his lifetime, largely because of a mutual friend, Gilbert White, an internationally respected geographer. I came to know White when he was a faculty member at the University of Chicago and later worked with him when he was chairman of the board of directors at Resources for the Future.

When I was at Resources for the Future, Kenneth Arrow came there from time to time to help with difficult problems. Arrow is an economists' economist. He has spent most of his career in the Economics Departments of Harvard and Stanford universities. I believe him to be the most profound economic theorist with whom I have associated. His discussions of traditional subjects have frequently opened my eyes to vistas I had not even imagined previously. He has authored more seminal publications than can be listed here. The Wikipedia entry on Kenneth Arrow and his faculty profile on the Stanford University website outline accomplishments and accolades that are indeed impressive and convey an impression of the role he has played in the economics profession.

Robert Solow has written extensively on a wide range of economic problems. Although he has been placed in the "equilibrate" column, he is perhaps best known for his investigation of the sources of economic growth in the United States. Traditional economic variables in his model failed to explain most of the growth that has been experienced; the unexplained portion was assigned to a residual category. Even though much of the economic growth was "unexplained," his model served the profession well for a long period. His economic growth accomplishments are distinguished from those of the new growth theorists by their efforts to bring economic growth determinants within an economic framework. As noted, Solow has made numerous contributions to economics in addition to economic growth studies.

The new growth theorists, such as Paul Romer and Paul Krugman, have attempted to provide economic theories of endogenous economic growth. They have done so by drawing on traditional, but often neglected, concepts in economics such as increasing returns and monopolistic competition (differentiated products). This has been combined with the study of economic institutions that will provide incentives for knowledge creation and application while providing for the benefits stemming from the potential public good effects of new technology (Warsh 2006). It would be premature to render judgment on the place of the new growth theory within the broad sweep of economic doctrine. It can, however, be said that it has resolved some of the paradoxes pertaining to comparative advantage among and within nations.

BACK TO KANSAS STATE

Dad died early in my time at Iowa State. I had just begun my most demanding course work in economic theory and statistics when I received a letter from Mom. That tough, self-reliant frontier woman was beside herself with anxiety. Dad was ill, and she was alone with him on the farm. After Merab and I read the letter, we jointly agreed I should go to the assistance of my mother. I arrived in the evening. Dad's first words to me: "You should not have left your studies to come home for me." We went to Independence, Kansas, the next day to see their physician. He provided medicine and instructions for Dad, and we started home. Almost immediately Dad suffered a heart attack, and I drove him directly to the emergency room of the hospital in Independence. I was alone with him at his bedside a short time later when he died.

After Dad's death, Mom decided she did not want to manage the farm. My younger brother and I stayed with her after the memorial services to hold a farm auction. I returned to Iowa State and, after missing more than a month of classes, found I was considerably behind my fellow graduate students.

After passing my preliminary examinations at Iowa State on the first try, I returned to K State in 1951 as an assistant professor.[7] I wrote my doctoral dissertation in absentia, a common practice then. It pertained to risk and uncertainty in wheat farming in western Kansas. I obtained historical data from a western Kansas branch experiment station that made it possible to test farmer strategies for coping with rainfall variation. This research established that the practices of farmers—practical people—dealt realistically with the risk and uncertainties inherent in Great Plains conditions (Castle 1954a, b, c).

As I completed work on my dissertation, I began to understand the meaning of scholarship and was far from satisfied with myself in that respect. I had taken a course in research methodology from Earl Heady when I was at Iowa State. I gained some appreciation in that course of how difficult it is to make a significant contribution to economic and social science knowledge. I was fortunate to have had an association with Earl Heady. I did not always agree with him, nor did I wish to follow in his footsteps. Yet he never failed to inspire, and he introduced me to subjects that made me recognize how little I knew. He possessed a fine intellect, worked exceedingly long hours, and traveled extensively both domestically and abroad. His scholarly output was prodigious. He published in the neighborhood of 800 journal articles and technical publications during his career. Kenneth Boulding once told me that Earl Heady was the only student he ever had who often wrote original economics as answers to course work examination questions.

In retrospect, the year at K State after I passed my preliminary examinations at Iowa State assisted my intellectual maturation considerably. I had gone to Iowa State to make myself acceptable for a faculty position at K State. I had accomplished that objective, but I was far from satisfied. I knew I was not an accomplished scholar, nor was I confident I knew how to become one. I believed I was a good, although probably not outstanding, teacher. Furthermore, there was a great deal of faculty controversy in the K State department then. I had not learned how to be a faculty colleague and at the same time provide for my own scholarly growth. I suspected several of the senior faculty (not including Ed Bagley) still thought of me as a graduate student. I resented their attitude, although I probably was more critical of myself than they were of me. I wanted to be useful in society either as a teacher, an advisor to practical decision makers, or a policy analyst, but I did not believe I was there yet. I was unable to decide how many of my shortcomings were because of me or my working environment.

THE FEDERAL RESERVE BANK OF KANSAS CITY

During the year I was working on my dissertation, Ray Doll, a professor at K State, resigned to accept a position at the Federal Reserve Bank of Kansas City. This was considered a major loss to the department. Ray was an excellent teacher who had specialized in agricultural policy. Soon after he left, he recommended to Clarence Tow, vice president of research at the Kansas City Fed, that I be considered for a vacant position there. Tow was assembling one of the better economic research departments within the Federal Reserve Bank system. His department consisted of three sections—financial, industrial, and agricultural economics. I was interviewed for, and subsequently accepted, an agricultural economics position under Doll. The economists there at that time held terminal degrees from Harvard, the University of Michigan, and the University of Minnesota. The discussions with the other economists were exceedingly stimulating. I learned a great deal about macroeconomics especially, because of its relevance to monetary policy. It was a great confidence builder to learn I could hold my own in these discussions. It also caused me to better appreciate my graduate education at Iowa State and Kansas State.[8]

The research department was responsible for providing intelligence to the Federal Reserve Bank of Kansas City and member banks in the 10th Federal Reserve District. We relied heavily on secondary data and subjective opinions of people throughout the district. This was useful experience. I concluded that the leading farmers with whom I had worked at Iowa State and K State made economic decisions that were just as good as, or better than, those of the better bankers with whom they did business. I also discovered that economics provided a perspective and a framework for analyzing information that people without such training did not have. We discharged our responsibilities with numerous articles and briefing papers, speaking engagements, and small group meetings and conferences.

With the benefit of my new perspective, I came to view the work of an agricultural economist at a land grant university differently. Opportunities occurred to me for novel applications of economic theory in teaching and research. Further, I missed contact with students and young people so much that I even taught Sunday school for a while. At the Fed, I was exposed to affluence on a scale I had previously only imagined. Rank, title, and status were manifested in both subtle and explicit ways. Allen Kneese, who became an economist at the Kansas City Fed after I left, told me a story that illustrated the culture differences between traditional banker types and former academics. The story pertained to regularly conducted bank tours and the economic research department with offices on the ninth floor of our building. During one such tour, a visitor inquired: "Do you have economists at the Federal Reserve Bank of Kansas City?" The tour guide, thinking the visitor had said, "Do you have *communists* at the Federal Reserve Bank of Kansas City?" replied, "No, we do not have any of those."

The tourist then said, "Oh, I thought there were some on the ninth floor."

At this the tour guide thought for a moment and then responded, "No, we do not have any, but if we did, that is where they would be, on the ninth floor!"

When I began to think about a possible return to academic work, I recognized that timing was of great importance. I did not want to make a change so soon that it created an impression I was not capable of working steadily in any position. On the other side of the equation, I realized that the longer I stayed at the Fed, the greater would be the salary reduction I would have to take when a move was made. Even though I knew the Fed experience was valuable, I also knew it would not be highly valued in academia. In the summer of 1953, I attended the annual meeting of the American Agricultural Economics Association held at Oregon State University in Corvallis, Oregon. While there, I met G. Burton Wood, head of the Department of Agricultural Economics at Oregon State. I expressed interest in an academic position that was available in his department. When I returned to Kansas City, I canvassed other land grant institutions in the nation. I was not overwhelmed by institutions seeking my services, but offers were forthcoming from Oregon State and Texas A & M. I joined the Oregon State University (then College) faculty July 1, 1954, as assistant professor, Department of Agricultural Economics.[9]

REFLECTIONS

I was 31 years old when I went to Oregon State in 1954. Approximately eight years earlier, I had resumed my college education at K State. I matured considerably during those eight years, both as a person and as an economist. I was more self-confident and, having obtained three economics degrees, I was a more accomplished economist. I wanted to be useful to practical people, and I had specific ideas how I could use my tools to do so. Furthermore, I wanted to work with young people and do what I could to help them realize their potential.

I was conscious of at least some of my limitations. I thought then, and continue to believe, that my principal weakness as an economist stemmed from my lack of formal training in mathematics. This has constrained my mastery of both economic and statistical theory. My aptitude tests have shown considerable capacity to think abstractly and to deal with symbols. Yet doing so has never appealed to me intellectually unless a solution to a practical problem required that I do so.

When the preceding is considered in a historical context, a pattern emerges. Typhoid fever in my junior year in high school prevented my taking advanced algebra at that time. That would have been corrected when I was in the Armed Forces if I had not taken myself from the specialized training program where I would have studied engineering. Finally, the events associated with my father's death resulted in my not taking more mathematically oriented courses when I was in graduate school. Thus, I could attribute my lack of formal mathematical education to a series of coincidences. Yet to do so would be a rationalization. If I had really desired to become more proficient in mathematics, I could have done so. Clearly, I had the ability, but I lacked the desire. I attribute my mathematical deficiencies to inadequate self-discipline. Even so, I have read a considerable amount of mathematically formulated economic theory. I find economic theory interesting and challenging, whether it is stated symbolically or verbally. (Anyway, are words not symbols?) I could speculate that had I become preoccupied with economics and quantitative analysis, I would not have acquired the knowledge in the philosophy of science and economic research methodology that I have. This, too, would be a rationalization. My methodological capacity would be greater if I were more accomplished in mathematics.

This mathematically deficient economist has no wish to dispense with mathematics in economics. Much of the progress economics has made would not have occurred without mathematics, in my opinion. Yet it would be unfortunate if economics were to become no more than applied mathematics (Rosenberg 1992). Some of the most important problems in economics were first uncovered in the absence of mathematics, and economic issues that are not easily or conveniently expressed mathematically have been abandoned or neglected in research and textbooks.

When I decided to seek academic employment while at the Kansas City Fed, I rather automatically began my search for an opening at land grant universities. I have asked myself subsequently why I did not also investigate departments of economics elsewhere in academia. There probably were two reasons. One was that I was not content to work only with economic theory *or* do application alone. I found the greatest satisfaction in moving from one to another. I was then, and continue to be, convinced this is a most appropriate way for applied economics to be done. The other reason was that I had a continuing interest in, and some knowledge about, agriculture, rural studies, and natural resources. I thought I would be more likely to capitalize on such knowledge in a land grant university than elsewhere in academia. Please recall, this was many years prior to Earth Day; the environmental movement had not yet materialized.

I was informed about the history and characteristics of existing programs in agricultural economics at land grant universities. For example, I knew about institutional economics centered at the University of Wisconsin. I also was aware of traditional farm management work at Cornell, Purdue, and the University of Illinois. I had long had an interest in farm management, but not traditional farm management that made little use of economic theory or decision theory. Even so, I was impressed by the influence three liberal arts institutions were having on agricultural economics—the University of Chicago with Theodore Schultz and D. Gale Johnson; Harvard University with John D. Black and John Kenneth Galbraith; and the Food Research Institute at Stanford with a number of well-published scholars. (None of these institutions now has a significant program in the economics of agriculture.) At that time, the greatest interest and prestige was accorded those doing work in agricultural policy, such as Schultz and Johnson at the University of Chicago; Galbraith and Black at Harvard; and O. B. Jessness at the University of Minnesota. I was intrigued with the question of how a small number of people at only a few places could accomplish so much.

When I left the Armed Forces in November 1945, and even for years thereafter, I thought of myself as unsettled and "footloose." By the time my family and I left the Federal Reserve Bank of Kansas City for the great Northwest, such feelings had largely dissipated. I was far from fulfilled professionally and did not know how Oregon State might work out, but I had definite ideas about who I was in a professional sense and what I wished to accomplish.

4

OREGON STATE
(1954–1975)

HOME AT LAST

I felt at home immediately upon undertaking duties and responsibilities at Oregon State University (then Oregon State College). Colleagues welcomed me, and administrators assigned me significant responsibilities. My family and I were nearly overcome with the beauty of Oregon. Many of the anxieties arising in other work experiences disappeared. Prior to moving to Oregon State, I had begun to wonder if I could be satisfied in any organization for a significant period.

Oregon State gave me considerable freedom in the discharge of my assigned responsibilities. Communication across academic units at OSU at that time was encouraged and not difficult. There were numerous avenues for promotion, tenure, and salary improvement. I published regularly in those early years, but I never felt pressure to do so. Over the years, there have been opportunities at more ideally structured, financially lucrative, and academically prestigious universities, but I am happy Oregon State has been my academic home for most of my intellectual journey.

The one serious anxiety I experienced early on at Oregon State stemmed from the administrative separation of economics from agricultural economics. The Economics Department was in Liberal Arts, at that time referred to as Lower Division. Liberal Arts departments were not permitted to have majors and could not offer disciplinary degrees. Their mission was to serve the remainder of the campus with undergraduate instruction. This was not good news in agricultural economics; our graduate students needed to take advanced courses in economics. In addition to these administrative problems and obstacles, for the first time I became aware of differences in traditions and aspirations between agricultural economics and economics. The administrative and cultural differences between the two have constituted the most serious professional problem I have had at Oregon State. I have been a part of several efforts to improve conditions, but the basic problem remains. The problem has two dimensions. One stems from an inevitable tension arising in many academic disciplines between those who are concerned mainly with theoretical or fundamental issues and those whose interest is doing useful work in society. This tension is explored in greater depth later, but it is not unique to Oregon State. The other dimension arises from conditions that are unique, or nearly so, to Oregon State, and it deserves brief treatment here.

During the 1930s, Oregon became one of the first states to create a state system of higher education. At that time, the two principal state-supported institutions of higher education were the University of Oregon and Oregon State University. The political powers in Oregon came to the conclusion that the best use of financial resources for the support of higher education required that duplication should be eliminated insofar as it was possible to do so. It was both possible and feasible to allocate certain professional fields. For example, the awarding of degrees in law and journalism could be made the responsibility of the University of Oregon. In a similar fashion, agriculture, forestry, and engineering could be vested with Oregon State. The fundamental problem that then arose was what to do about the arts and sciences. Obviously, any person who aspired to be well educated would not wish to be denied access to basic arts and sciences. Further, it was clear there was need for excellent science in several professional fields at both the University of Oregon and Oregon State. But the same argument was not believed to apply to the liberal arts and the social sciences at Oregon State. Why was it necessary that students in agriculture, forestry, engineering, home economics, and pharmacy, for example, be taught by faculty also offering disciplinary degrees in such fields as English, sociology, music, and psychology? It was from such thinking that the Lower Division (Liberal Arts) at Oregon State arose. The enforcement of curricular allocation in the State System of Higher Education in Oregon became the responsibility of the chancellor's office, located on the University of Oregon campus in Eugene, Oregon.

When I later served as Oregon State's dean of faculty, and then as dean of the Graduate School, I was much impressed by the rigidity, but frequent internal inconsistency, of the allocation enforcement system. By that time, there were four regional colleges that had become a part of the State System of Higher Education. I discovered, by accident, that these regional colleges had been granted greater freedom in their graduate liberal arts offerings than was permitted at Oregon State. Perhaps this came about because of the fierce competition between Oregon State and the U of O. Or perhaps it was thought that if there was suspicion that the allocation system was not working, evidence of duplication most likely would be found at the University of Oregon and Oregon State. Oregon State University cannot be understood without knowledge of the historic curricular allocation system, although it is now much less constraining and rigid than when I arrived in 1954. I will return to this issue later in the book.

The standing Oregon State enjoyed with Oregon citizens in 1954 was higher than what I had observed at most institutions in the Midwest. The natural resources and agriculture of Oregon were, of course, very different from those with which I was most familiar. The opportunities for interesting investigations were nearly overwhelming. I asked to work on a few smaller problems in different parts of Oregon. One such problem existed in the Christmas Valley of south central Oregon. A land developer from California had settled there and acquired options on considerable land, much of it abandoned by earlier settlers. The developer had drilled irrigation wells and advertised that with modern technology it was possible to grow several crops profitably. Both the Soil Conservation Service (SCS) and the Farmers Home Administration (FHA) wanted Oregon State to become involved. The FHA was getting requests for loans they doubted could be repaid. The SCS was being

The Oregonian

ESTABLISHED BY HENRY L. PITTOCK
An Independent Republican Newspaper
Published Daily, except Sunday, by The Oregonian Publishing Company, Oregonian Bldg., 1320 S.
Broadway, Portland 1, Oregon, which also publishes The Sunday Oregonian, Telephone CA 6-2
MICHAEL J. FREY, President and Publisher
HERBERT LUNDY, Editor of the Editorial Page   ROBERT C. NOTSON, Managing Edi
HAROLD V. MANZER, Advertising Director   LEWIS J. CASCADDEN, Circulation Mana

SATURDAY, MARCH 17, 1956   3M

Fort Rock Censorship Try

An agricultural experiment station is not a booster club. Its duty is to publish the bad along with the good in any investigation of an agricultural problem or situation. Unless its reports can be accepted as entirely frank, they have no value.

That is the basic point raised in the suit brought by residents of the Fort Rock basin in Lake county to enjoin the agricultural experiment station at Oregon State College from publishing further articles about farming possibilities in the basin. The plaintiff, backed by nearly all the ranchers in the area, charges that an article in the winter edition of the quarterly, "Oregon's Agricultural Progress," was inaccurate and detrimental to development of the basin. The article was based on studies made by OSC and Soil Conservation Service agricultural economists.

"Unpredictable—usually short—growing seasons, inability to get allotments on publicly owned grazing lands, and long distances to market will combine to squeeze new irrigation farmers," said the article. "Needed for success: plenty of money and good management. Even then, things will be tough . . ." Irrigation of frost-hardy crops such as alfalfa is the only way of insuring some measure of success, said a footnote.

Such a view from an official source was a blow, of course, to those who have brought about construction of a $1,000,000 power line into the basin to pump irrigation water from

an underground lake. Some of them had forecast cash crops growing on up to 100,000 acres of irrigated land. We are in no position to judge which view is correct, but, unless bias can be shown, the experts were obliged to release their conclusions arrived at by scientific analysis.

Fort Rock was fairly thickly settled once before by homesteaders, most of whom found the going too tough. Had there been such investigations then, the early settlers might have been spared failure. Success of the Fort Rock development cannot be attained by censoring criticism but by proceeding on sound principles. If the experiment station experts are wrong, they can be proved so by equally authoritative sources. If they are silenced in the future, except when they have something favorable to say, they may as well close up shop. No one will have any faith in their statements.

---

asked if the soils would accommodate irrigation, but SCS personnel were convinced that soils were not the most limiting factor. The short growing season was more limiting than either soils or the remote location.

I agreed to lead an investigation of the feasibility of irrigation in Christmas Valley. The SCS assigned a senior economist, Carroll Dwyer, to work with me. We came to pessimistic conclusions about irrigated farming there, except for extensive crops such as alfalfa hay. The developer learned of our conclusions when they were in manuscript form and brought suit against Oregon State to stop publication. Here I was, the first crack out of the box, an assistant professor without tenure, bringing a lawsuit down on my university!

This experience taught me how responsible administrators behave under pressure. A. L. Strand, OSU president, told F. E. Price, dean of the School of Agriculture, that he, the president, had another difficult problem then and that the dean should handle Christmas Valley. Price then went over our manuscript with me sentence by sentence, asking me to defend each one. I defended every sentence save one. The dean told me to reflect on the one I could not defend. I did so, and changed the sentence, although our conclusions were not affected. From that point on, Price never wavered in his support. An Oregon assistant attorney general defended Oregon State in the lawsuit. The plaintiff's motion was denied on grounds the suit was incorrectly filed, and we thought we were home free. However the developer persuaded several of his farmer neighbors to go with him to see the governor. The governor directed the group to go to Oregon State to discuss the matter. At about this time, the *Oregonian*, the state's leading newspaper, published an editorial that opened with the sentence, "An agricultural experiment station is not a booster club." This was good news for me, but the flap was not over yet.

The developer and his group did come to Oregon State for a meeting. The dean presided, they presented their complaints, and I responded. Except for the developer, most of the group seemed convinced by my arguments. As these events were unfolding, the lawsuit was dismissed from court a second time on the grounds of freedom of the press. It was a relief to have this over with, and I learned much from this minor scrimmage (Castle and Dwyer 1956).

The assistant attorney general was somewhat disappointed by the outcome. He said, "This is such an interesting case; I would have liked to argue it before the Oregon Supreme Court."

I was less interested than he in establishing legal precedent.

THREE SIGNIFICANT RESPONSIBILITIES

Soon after I arrived at Oregon State, I was assigned three responsibilities: 1) provide instruction for an undergraduate course in farm management and a graduate course in production economics; 2) develop a research and public education program in water resource economics; and 3) provide leadership for the department's graduate program. There was an active masters program and authorization to offer the doctorate.

TEACHING FARM MANAGEMENT: My farm management teaching responsibility led me to write *Farm Business Management*, a textbook designed for sophomores and juniors. For several years, Manning Becker and I taught sections of an undergraduate course in farm management. Manning also had an Extension responsibility that I helped him discharge, making it possible to obtain information from farmers. I learned that farmers approached many problems differently than one would conclude from the way most farm management books were organized at that time. As a consequence, the textbook I decided to write was organized to be consistent with the way farmers viewed problems. Manning Becker prepared three chapters pertaining to farm records for the textbook.

I established three guidelines for the text:

1) It would be accessible to students without previous training in economics or accounting.

2) The economics used would be consistent with state-of-the-art economic theory of the firm (the basic economic organization of businesses constituting an industry).

3) It would reflect how farmers and farm managers view the problems they face and the decisions they make.

There have since been two revisions—one with Fred Smith (1972) and one with Gene Nelson (1987), both colleagues in my department at Oregon State. It has been in print approximately four decades and has been translated into (I believe) six languages. More than 100,000 copies have been sold. Judged by traditional standards, it has been a success. I make no apologies for the elementary nature of the material, but I am uncomfortable with my acceptance of some conventional wisdom pertaining to neoclassical economics as it existed at that time. For example, the conventional wisdom of the time was that constant returns to scale existed in farming (an increase in the factors of production resulted in a proportionate increase in output). This supposedly provided a rationale for farms of different size existing side by side in American agriculture. Yet it was obvious that farms had become larger and farmers had declined in number with the passage of time. We had a responsibility to note this inconsistency and advance possible explanations. I regret that I did not do this in *Farm Business Management.* Much agricultural policy literature suffers from a similar omission.

A WATER RESOURCES ECONOMIC RESEARCH PROGRAM: The development of a water resources economic research program became a most significant part of my intellectual journey (Castle 2000). In the 1950s, regional committees were financed and sponsored by the Farm Foundation to facilitate and encourage cooperation among land grant university economists on important regional problems. I became Oregon State's representative to the Western Water Resources Research Committee. Meetings emphasized needed research on water programs and institutional public policy. The meetings were multidisciplinary in nature, with political scientists, lawyers, and engineers, as well as economists, in attendance. Lawyers and engineers had been the most important professionals in western water development, but it was clear that additional expertise was required.

The committee meetings were among the most intellectually stimulating experiences I have had. The policy issues pertaining to natural resources ("land," in classical economics) involved rules, customs, fragmentation, multiple jurisdictions, and social conflict. Yet the production economics research and teaching at Iowa State, based on neoclassical economics, considered natural resources to be no different than any other economic asset. Land economists were sometimes referred to as "factor specialists," suggesting they were not connected to mainstream economics, even though Adam Smith and Alfred Marshall had devoted considerable attention to institutional issues pertaining to land. The water committee members did not get diverted by matters of this nature. They just took water problems in whatever direction seemed promising.

Two committee members made contributions that influenced my entire career. One was S. V. Ciriacy-Wantrup. His book on conservation was first published in 1952 and has become a classic. His "safe minimum standard" continues to guide public policy, and it was he who first called attention to the possible use of questionnaires to ascertain values people place on environmental quality (Ciriacy-Wantrup 1952). The contingent valuation methodology that has received so much attention in recent years stemmed from this suggestion. The other person was Mark Regan. He was the United States Department of Agriculture's continuing member of the federal interagency group that created the "Greenbook" (Report of the Interagency Committee). This provided the framework for applying benefit-cost analysis to water resource development projects. As an employee of an agency contributing to the interagency group, Mark was not at liberty to publish independently on those issues and discussions. Nevertheless, he was exceedingly generous with his time and frank in his discussions with the water committee. He understood the technicalities of welfare economics and correctly anticipated benefit-cost and water resource issues that dominated the water resources literature for several years (Regan 1964).[1]

The Oregon State water economics program that I led placed emphasis on the economic comparability of market and extra-market values in water use and allocation. Outdoor recreation and water quality, in particular, received attention. Clifford Hildreth and Leonid Hurwicz, both economists on the faculty of the University of Minnesota, and Allen Kneese, an economist with Resources for the Future (RFF), came to OSU to assist with the design of specific projects. During this period, I was fortunate to develop a lifelong association with Charles Warren, a fisheries biologist in the OSU Department of Fisheries and Wildlife. Long before it became accepted in resource and environmental economics, he taught me that a person may derive satisfaction from knowing something exists in nature, even though they do not experience it directly and individually. This experience has since come to be labeled "passive use values" and "existence values" in economic literature. The estimation and consideration of such values now constitutes a significant economic literature.

Charles Warren also called my attention to the political significance of the relative economic value of the sport and commercial fishery in Oregon. He suggested I inquire of the Oregon Game Commission if they would finance an economic research study of the salmon-steelhead sport fishery. At that time, a small group of economists, myself included, were working on conceptual models that would permit the economic value of outdoor recreation to be brought into an economic framework. Marion Clawson, at RFF, came into possession of a one-page letter that noted economist Harold Hotelling had written to the National Park Service in 1949. In this letter, Hotelling stated how travel expenditures for outdoor recreation purposes could be used to estimate a demand function for outdoor recreation. Clawson, using hypothetical data organized as suggested by Hotelling, illustrated a demand function in a lecture given at the University of Wisconsin in 1959 (Clawson 1981). I decided to use the Hotelling-Clawson model to analyze data that would be probably the first empirical estimate of a demand function using the latest econometric techniques.

The Oregon Game Commission made the estimate of the demand function of the salmon-steelhead fishery possible not only through financing, but also by providing crucial data. The Game Commission maintained data on expenditures by salmon and steelhead sport fishers who resided at different distances from their fisheries. I enlisted the assistance of William Brown, a longtime colleague who was more accomplished in econometrics than I, as well as Ajmer Singh, a graduate student. The publication resulting from this research has been cited extensively in the literature and has attracted the attention of policy analysts (Brown, Singh, and Castle 1964).

Work with the Water Resources Research Committee experience, welfare economic theory, and agency people in the Pacific Northwest resulted in an evaluation of benefit-cost analysis in both theory and practice (Castle, Kelso, and Gardner 1963). The amount of data required for careful benefit-cost analysis is considerable. The results are sensitive to particular parameters, as skillful analysts soon recognize. Research economists, relying on ex post facto (after the fact) analysis of actual projects, typically have been critical of ex ante estimates made by agency economists. Yet few research economists have made, and then compared, their own predictions with actual results for real-world projects.

Charles Warren was also instrumental in helping me initiate research on the economics of environmental quality. He called my attention to a water quality problem in Yaquina Bay, the estuary touching both Toledo and Newport, Oregon. Georgia-Pacific, an integrated forest products firm, was then locating a pulp and paper plant at Toledo, near the east end of Yaquina Bay. There were questions regarding how the effluent from the plant was to be treated if environmental quality of the estuary was to be protected. An economically and environmentally valuable recreational activity would be negatively affected if untreated effluent were discharged into the estuary. I prepared a proposal for a study of the economics of environmental quality and outdoor recreation in Yaquina

Herbert Stoevener

Bay and submitted it to the federal agencies that were the predecessors of the Environmental Protection Agency, and was funded by them. I have been told ours was the second grant made by a federal agency for such a study. Such investigations are now commonplace.

Herbert Stoevener came to Oregon State for postdoctoral work as my research associate and managed the investigation in cooperation with the departments of Fisheries and Wildlife and Civil Engineering. We did considerable work applying state-of-the-art quantitative methods in the economic evaluation of nonmarket phenomena. In particular, the dissertation of Joe B. Stevens, one of my doctoral students, was an original examination of several economic relationships associated with the estuary and its use (Stevens 1965, 1966). The study also directed attention to fundamental questions about the relation of national benefit-cost analysis to regional (or intermediate) decision making. If national benefit-cost makes no allowance for variation in regional comparative advantage, and if local or regional institutions are bound by traditional national economic efficiency measures, both environmental quality and economic welfare may suffer. The policy implications of the Yaquina Bay and similar studies continue to find their way into the literature (Stoevener et al. 1972; Castle 2008). The comparative advantage of a particular place or community may be natural, human, or community-based. Incorrect perceptions of place-based comparative advantage lead to flawed public policy: Underestimation will lead to the loss of community income or the sacrifice of environmental values in pursuit of less valuable economic development. Overestimation will result in counter-productive efforts to attract economic development. Both the American Agricultural Economics Association and the Western Agricultural Economics Association subsequently cited the Yaquina Bay investigation for research excellence.

GRADUATE EDUCATION IN THE DEPARTMENT OF AGRICULTURAL ECONOMICS: Shortly after I came to Oregon State, I was assigned leadership of graduate education in the Department of Agricultural Economics. My immediate assessment was that we could offer an outstanding masters degree, but we were not staffed adequately to offer a high-quality doctoral program. Federal government policies at that time created incentives for academic institutions to offer the doctorate; a doctorate-granting institution had better credentials for obtaining training grants and research awards than did those that offered the masters as its terminal degree. This may have influenced OSU administrators in favor of offering a doctorate in agricultural economics. I had a voice in faculty appointments and financial assistance for awards to graduate students, and I had developed relationships with cooperating departments.

Although offering the doctorate was not my decision to make, I concluded there were steps that must be taken if the degree were to have some degree of quality. I established the following guidelines:

- An agricultural economist must be a good economist

- Graduate students should master economic doctrine at a level appropriate to the degree they seek. Dissertation research will address problems that require economics as a tool of analysis.

- Competence in the neoclassical core of economics will be expected of all students, but perspective is desirable as provided by the study of economic doctrine and research methodology.

- Major professors and graduate committee members will be available for consultation when graduate students conduct their dissertation research. (Faculty availability was a major problem with some prestigious graduate programs at that time. I concluded that faculty availability could be a partial substitute for reputation.)

- No attempt would be made to treat graduate students as if they were "peas in a pod," or nearly identical. Graduate faculty in the departments of Agricultural Economics and Economics represented a pluralistic methodology, having received diverse education from several universities.

In retrospect, I was, and remain, impressed with how rapidly the Oregon State graduate program in agricultural economics began to command respect, first across the United States and then internationally. For example, several masters graduates from that era achieved significant professional recognition. OSU masters graduates Paul Barkley (now a fellow of the American Agricultural Economics Association), L. T. Wallace, Randolph Barker, LeRoy Rogers, and Frank Conklin went on to obtain doctorates at other institutions and then established reputations in agricultural economics. Michael Nelson, one of the first doctoral students in the Oregon State program, received an outstanding thesis award for his dissertation. He subsequently authored a Resources for the Future book. Two other early doctorate recipients—Harvey Hutchings and Virgil Norton—had noteworthy careers in academia and with the National Marine Fisheries Service.

During the decade of the 1960s, environmental concern became of increasing importance nationally, culminating in Earth Day in 1970. Fortunately, a number of scholarly efforts concerned with this subject had been underway for some time. Resources for the Future (RFF) came into existence in 1952 and contributed to a growing and valuable body of literature on the subject. The RFF publications provided great support for the few graduate programs emphasizing resource and environmental economics that were scattered across the country. By the time of Earth Day in 1970, a significant cadre of resource and environmental economists was deployed across the country, able to contribute to various facets of public policy.

During the 1950s and 1960s, the Oregon State Department of Agricultural Economics was one of a limited number of places where a student could specialize in resource and environmental economics. In the summer of 1969, the department offered a special summer session in resource and environmental economics. John Krutilla, Resources for the Future, and Chester Baker, University of Illinois, were brought to Oregon State as faculty. The session was advertised widely, and graduate students as well as mature scholars from across the nation attended.

Early in the 1970s, the name of the Oregon State Department of Agricultural Economics was changed to Agricultural and Resource Economics. Outstanding graduate students came to the department to study both before and after Earth Day, providing support for the contention that young people have the capacity to identify and anticipate emerging developments within a field of study. Six graduates of the department from that era became fellows of the American Agricultural Economics Association, the highest honor the Association bestows. One, Alan Randall, has received an enduring publication award from both the American Agricultural Economics Association (now the Agricultural and Applied Economics Association) and the Association of Environmental and Resource Economists. Another, Sandra Batie, was the first woman to be elected president of the American Agricultural Economics Association and the second woman to become a Fellow of that Association. Another, Daniel Bromley, is among the 1,000 most-cited authors in the entire economics profession. Still another, Bruce Beattie, also had the distinction of serving as president and being named a fellow of the American Agricultural Economics Association.

It is my view that the faculty in economics at Oregon State from 1955 to 1972 have not received appropriate credit in the profession for their contribution to graduate education. In agricultural economics, a partial list includes William Brown, John Edwards, Albert Halter, Richard Johnston, Bruce Rettig, Joe Stevens, and Herbert Stoevener. In the Economics Department, Charles Friday, Charles Vars, John Farrell, and Bill Wilkins deserve mention. A diverse group often can accomplish great things by cooperative effort, if they are motivated by a common objective. The momentum of these early years continued for several additional years, and OSU graduates continued to receive professional recognition. One, David Ervin, returned to Oregon State to become department head for a period.

ACADEMIC ADMINISTRATION[2]

As the decade of the 1960s unfolded, I was aware that personal opportunities probably would arise in academic administration. I enjoyed decision making and the analysis of difficult problems. My greatest professional satisfaction came from contributing to the professional development and recognition of students and younger colleagues. I recognized that an administrative position might provide greater leverage for accomplishments of this nature than the faculty position I had been occupying. Even so, at that time I had no intention of leaving my chosen discipline of economics. I came to believe that the position of department head or chair would provide opportunities to influence both economics and economists.

In 1965 I accepted an invitation from Washington State University to interview for chair of Agricultural Economics. The evening before my wife, daughter, and I were to leave for Pullman, James Jensen, president at Oregon State, called and asked that I come to his home, and of course I did so. Jim Jensen and I did not know one another well at that time. I had served on the Oregon State Faculty Senate Executive Committee the previous year, and we first came to know one another slightly as a consequence of that.

Two years earlier, the dean of faculty position had been created at Oregon State. Until then, there were only two campus-wide academic administrators with the rank of dean or above—the president and the dean of administration. The dean of faculty was to be the principal academic officer of the University with responsibility for academic personnel, curricular affairs, and faculty governance. The incumbent had been in the position for approximately a year, but he and the position he occupied were not well suited to one another. James Jensen told me the position would soon be vacant, and he would like me to occupy it. What a shock! I had never considered surrendering my disciplinary work in economics for academic administration. I preferred living in Corvallis rather than Pullman, but academic department administration was more attractive than a campus-wide responsibility.

Jensen was offering the position to a person without previous administrative experience, in effect elevating me over the heads or department chairs and deans then on campus.[3] My family and I made the trip to Pullman as planned, and I was offered the position there. After I returned to Corvallis, I declined the Washington State offer and accepted the invitation to become dean of faculty. I am not prone to depression, but I was more discouraged during the month following the decision to accept than before or since. The discouragement I experienced stemmed from the fear that, as an administrator at that level, I was likely to lose competence in economics and standing in the economics profession. Fortunately, I soon became very busy. The incumbent left several unfinished tasks, and I became heir to a major responsibility.[4] I acquired confidence in my ability to be dean of faculty and developed much respect for President Jensen, as well as Milosh Popovich, dean of administration. Both became lifelong friends.

As my first year as dean of faculty came to a close, numerous administrative changes occurred in the College of Agricultural Sciences. The position of director of the Oregon Agricultural Experiment Station became vacant and was offered to and accepted by G. Burton Wood, head of Agricultural Economics. I requested permission from James Jensen to apply for head of Agricultural Economics. My vacating the dean of faculty position would disrupt Jensen's plans for the University, but he knew of the pain I experienced when I became dean of faculty. Jim Jensen, a big and generous person, granted my request, and I became head, Department of Agricultural Economics, July 1, 1966. I experience both gratitude and guilt when I reflect on Jim Jensen's generosity. It was only with the passage of time that I recognized how awkward I made it for him by resigning as dean of faculty.

My six years as department head were the most satisfying of my professional career. The experience as dean of faculty was exceedingly valuable to me as a department head. The department had competent faculty, and I was able to make several young additions. We reorganized the decision structure and teaching assignments for the department. I continued to teach and write in my chosen field. There were numerous campus-wide developments that provided opportunities for the department, and we took advantage of as many as possible. Oregon State became one of the nation's first Sea Grant Colleges during this period, with an economics component administered from our department.

Several department activities were innovative, although we were only addressing basic problems in a logical way. Shortly after I became head, enough "soft money" was obtained to hold a faculty retreat on the Oregon coast. Strategic planning with an emphasis on "visioning" was becoming popular at that time. We made use of this methodology but avoided pitfalls that are now recognized.[5] We developed strategies that were specific as to how we would react to new developments, as well as to probable operations and projects. In particular, we relied heavily on participatory decision making and a long-range planning committee. This committee was given autonomy to bring new ideas, approaches, and opportunities to the department head. The head, in turn, was obligated to refer long-run issues to the committee for comment or recommendation prior to taking the matter to the entire department. Priorities became guides, but not dogma. Some of these innovations remain operative, having survived four decades and six department heads.

As the decade of the 1960s came to an end, James Jensen decided to resign. These were turbulent times in higher education generally, as well as at Oregon State. At that time, major academic units were labeled "schools," but some wished to become "colleges." Oceanography was a department and wished to become a school or college. The School of Science administration was challenging traditional administrative arrangements on several fronts. Numerous other questions existed pertaining to faculty and student affairs. Jensen was concerned about leaving the University with problems that would fester until his successor was chosen and became familiar with the place. His solution was an unorthodox one. He appointed a three-person University Goals Commission to study problems the University faced and make findings about reasonable aspirations for the University. The Goals Commission experience will appear again later in this book. It was a unique and stimulating intellectual activity. We did not have a chair and proceeded only by consensus. We examined every facet of the University and made recommendations pertaining to goals and objectives, as well as to operating procedures. The traditions and procedures of different disciplines and fields caused me to view economics in a broader context than I had prior to that time. The experience found its way into articles about interdisciplinary scholarship and University governance (Castle 1970, 1971).

Near the end of my tenure as department head, Roy Young, vice president for Research and Graduate Studies, assigned me responsibility to administer a major grant made to Oregon State by the Rockefeller Foundation. Oregon was justifiably recognized as an environmental leader, and there was interest in a systematic view of this experience. The focus of the project was the best use of Oregon's natural resources during periods of rapid social and economic change. This was a multidisciplinary undertaking and involved personnel from several schools and departments. The grant permitted several young scientists to join the faculty, including Bruce Shepard in Political Science, Owen Osborne in Engineering, and James Fitch in Agricultural Economics. Kenneth Godwin, Political Science, and George Carson, History, were involved as well. Some of the first writing of William Robbins, History, concerning Oregon natural resource history was made possible by the grant. The professional development of these people and numerous graduate students constituted a significant contribution of the grant. Policy briefs written as a result of the project anticipated national policy issues well, but it is less clear if they had significant effect on public policies in Oregon.

Demanding administrative assignments, academic or otherwise, often take a toll on family members other than the one with the assignment. Had I accepted the Washington State opportunity, it would have been necessary that my daughter change schools for her senior year of high school. My wife's health made social obligations difficult for her. I have long had concern about staying too long in one position. I believed I had made my major contribution to the Department of Agricultural Economics by the end of my sixth year as head. In retrospect, I doubt this should have been a concern.

In any event, I became dean of the Oregon State Graduate School in 1972. This time I was dean for three and one-half years, the average tenure for graduate school deans across the land. As graduate dean, I completed my administration of the Rockefeller project and was able to bring about some change in the Graduate School as well. The in-depth periodic reviews of department graduate programs by the Graduate School which I instituted have endured to the present.

An opportunity of a lifetime came my way in the spring of 1973, although negotiations had begun the previous summer. Three agricultural economists at the Ford Foundation knew me well—Lowell Hardin, Dale Hathaway, and Norman Collins. They said they were impressed by the program at Oregon State in resource and environmental economics, and they knew of my administration of the Rockefeller Foundation project as well. They approached me about spending time in India on behalf of the Ford Foundation. They wanted an assessment of India as a place for a Ford Foundation program in resource and environmental economics. Of course, I was interested, and, after acceptance, devoted approximately six weeks to the assignment. I needed to stop in Washington, D.C., on my return. As a result, the Ford Foundation arranged my itinerary so that I could travel around the world on the trip. They sent me first-class all the way, with stops in Tokyo, Hong Kong, Bangkok, New Delhi, Rome, London, and Washington, D.C. On the evening of April 13, 1973, on a sidewalk in Madras, India, I suddenly realized it was my 50th birthday.

REFLECTIONS

The time span covered by this chapter is 21-plus years. When I reflect on my intellectual journey over these years, I can hardly believe my good fortune. I arrived at Oregon State in 1954 with many doubts about whether what I had to offer would be acceptable in academia. Approximately 21 years later, I had the privilege of working with exceptional graduate students, recruiting promising young faculty, serving as academic officer for my university, and traveling around the world for a prestigious foundation. Surely there are lessons to be learned from such experiences.

Four subjects provide focus for reflections stemming from those 21 years. The first pertains to the way the philosophy of science, research methodology, and economic theory influenced my view of universities, academic administration, and graduate education. The second deals with a paradigm change in agricultural economics. This is followed by specific observations about academic administration. The fourth deals with graduate education.

A VIEW OF THE ACADEMY: When I returned to the Department of Agricultural and Resource Economics as department head in 1966, our faculty there took a fresh look at our course offerings, including who would teach particular courses. My teaching assignment included our graduate-level course in research methodology. This was not a course in research methods, but rather, it encouraged students to view economic theory through the lens provided by the philosophy of science. When I first began to teach research methodology, the emphasis was on the modifications Karl Popper had brought to logical positivism. The work of Thomas Kuhn then came on the scene and had an enormous impact on my views. Two facets of Kuhn's model of scientific revolutions stand out. One is the distinction he makes between revolutionary and normal science. The other pertains to his view of the organization of scientific endeavor. Each requires a brief explanation. (See also Appendix C.)

According to Kuhn, a development in science deserves to be called revolutionary if it is sufficiently novel and potentially productive to attract others to its elaboration. Kuhn calls such elaboration efforts "normal science," which he further breaks down into three categories. One is methods of study and instrumentation. A second is "puzzle solving" or theoretical work not made explicit by earlier revolutionary accomplishments. The third is making empirical observations that are made interesting by the discovery.

The above classification directs attention to how scientific endeavor is organized. Kuhn advances the notion of a "disciplinary matrix" that provides for important activities that stem from the original discovery. This includes the literature of the field, especially peer-reviewed journals and textbooks. It also provides for attracting new workers (graduate students and postdoctoral workers) into the field. The concept of a disciplinary matrix is helpful in understanding the activities within various organizational units (say, departments) in academic institutions.

Kuhn's view of science helped me understand why workers within an academic discipline typically are defensive of prevailing views within established fields. In contrast, Karl Popper, an earlier philosopher, believed the true scientific type would be actively searching for evidence that would overthrow a prevailing view. Yet Kuhn also advanced a model of scientific change consisting of two parts. One part involved the accumulation of anomalies—pieces of evidence or circumstances that existing theory does rationalize or explain, plus an alternative theory that rationalizes the anomalies as well as at least some of the reality explained by existing theory. I came to realize that Kuhn's views can be applied usefully to more than just scientific activity within academic institutions.[6]

I have also made use of economic theory when managing and evaluating academic endeavors. Microeconomic theory provides a rationale of how consumers and producers behave in markets. Yet higher education is generally regarded as operating in a nonmarket environment. How does one compare apples and oranges?

First, consider how markets achieve efficiency. One way is to reduce output to homogeneous units that can be created or processed from one blueprint. Think of tubes of toothpaste or Big Mac hamburgers. One might ask: "How can that be? Toothpaste and Big Macs are produced under monopolistic competition with heterogeneous products, not perfect competition with homogeneous products." The explanation is a simple one. The toothpaste or hamburger industry indeed produces differentiated products, but brands make it possible for producers to create the belief by consumers that, say, Big Macs are different from other hamburgers. To the extent this can be done, they become homogeneous products to the producers of Big Macs, and they make it possible to reap economies of scale. Yet education, certainly graduate education, must take individuals into consideration, and individuals are not homogeneous. The important step is to isolate those educational processes that can be made routine from those that should not be. In Endnote 4, an Oregon State activity is described that involved the selection of 64 outstanding teachers in less than one academic year. This was accomplished by standardizing criteria and the general process to be used among and across the individuals from the entire campus. Yet the final choices were of individuals. When I went to RFF, there was great concern about the amount of time that it took to conduct peer review and arrive at a consensus decision with regard to manuscripts submitted for publication. I was told this was because each manuscript was different and there was not much room for improvement. Subsequently, we reviewed this process, installed reforms, and began to review and render decisions in less than half the time that had previously been used. If efficiency gains can be achieved and internalized within a managerial unit, they may have the same effect as a budget increase.

A PARADIGM CHANGE IN ECONOMICS AND AGRICULTURAL ECONOMICS: With the benefit of hindsight, I now reflect that I have participated in a paradigm shift in agricultural economics during my professional career, beginning with my first tour of duty at Oregon State. I refer to the emergence of resource economics as field of specialization within economics and agricultural economics. It also involves my relationship with my major professor, Earl O. Heady, at Iowa State University.

As noted earlier, Earl Heady was a remarkable economist indeed. It is my opinion that his *Economics of Agricultural Production and Resource Use* is the most impressive publication ever written by an agricultural economist. It not only deals with subjects identified in the title of the book but also with much related material. The theoretical basis for much agricultural policy literature is set forth in his treatments of technical change and agricultural price formation. His command of neoclassical economic concepts permitted such discussions to be focused in a concise and rigorous fashion, in contrast to much of the related literature in that era.

As I began to organize a research program at Oregon State dealing with natural resource policy, I recognized that Heady's view of natural resources was not consistent with what I observed in reality. Heady followed the great economist Frank Knight in maintaining that the classic classification of the factors of production—land, labor, and capital—could be collapsed to two—labor and capital. (The classical economists used "land" to encompass all of the beneficial powers of nature.) Knight, and then Heady, denied there was anything sufficiently novel or different about land to justify separating it from other forms of capital. My observation and reflection as I was developing a research program was that society did, in fact, judge that land—or natural resources or the natural environment—was different. Some land was not priced in the market, and when it was, it was often subject to institutional constraints far different from other forms of capital. I then concluded that the writings of Ciriacy-Wantrup would be a better base for research in resource economics than Heady's approach. Ciriacy-Wantrup's entire orientation was based on the conviction that natural resources, or "land," was different from other forms of capital.

At that time, it did not occur to me that my decision was a part of something bigger—a paradigm shift. It just seemed logical to me as a way to proceed. Kuhn's work in the philosophy of science was not generally known then. Beyond that, no one then could foresee the difficulties economic theory would have in accommodating the complexities that have arisen in integrating the natural environment into the corpus of economics as a discipline.

As I review the emergence of resource and environmental economics after I came to Oregon State in 1954, I am impressed by the number of parallels with Kuhn's model of paradigm changes. Yet there was one important difference. Kuhn was concerned with basic science, mainly physics and chemistry, not applied science. Resource and environmental economics is applied science. It was not an accident that it achieved its autonomy within economics and agricultural economics after Earth Day in 1970.

Paradigm shifts often occur over fairly long periods, and that has been the case with resource and environmental economics. The first move was made in the 1950s, when the early resource economists resurrected the classical economists' classification of the factors of production as land, labor, and capital. In doing so, they rejected the Knight and Heady position that land was just a form of capital and that only labor and capital were necessary. This did not mean that Knight and Heady were necessarily incorrect in using only two factors of production for certain purposes. It was necessary to consider land as different only when natural resources became the focal point of investigations. In this respect, resource economists were no different than land economists who had specialized in the study of land-related institutions. Resource economists were different from land economists, however, in the extent to which they made use of neoclassical economic theory. Neoclassical economics, for example, provides a theoretical base for benefit-cost analysis and externalities (Castle 1965).

The early resource economists were concerned with the use of natural resources in the production of commodities routinely priced in the market, as well as those that were not. A major early research thrust was concerned with non-market evaluation, especially the non-market valuation of water quality and outdoor recreation. Wantrup had noted that people could value resources for their own sake and not just for the contribution they made to the creation of commodities. In 1954 and 1955, Paul Samuelson developed the pure theory of public goods, of obvious relevance to certain natural resources. In 1967, John Krutilla, an early resource economist, provided a systematic treatment of why economic analysis should consider resources as being of value for their own sake.

In April 1970, the first Earth Day occurred. This signaled that societal attitudes about the natural environment were changing, and resource economics was affected accordingly. The Kuhn model of paradigm shifts does not consider public attitude; rather, it is confined to scientific affairs. An applied science, such as resource economics, derives its problems from society. And Earth Day was fundamentally concerned with resources for their own sake, not just for their contribution to the production of commodities. Public funds became available for economic research, and university departments of economics and agricultural economics began to recognize resource economics as a graduate study specialty. Many agricultural economics departments began to change their names to Departments of Agricultural and Resource Economics or Agricultural and Applied Economics. "Environmental Economics" was substituted for "Resource Economics" in some cases, and "Resource and Environmental Economics" in others.

Without the significant public interest in the natural environment that arose in the 1970s, it is doubtful resource economics would have the standing in economics and agricultural economics it now enjoys. Prior to Earth Day, the American Agricultural Economics Association (AAEA) questioned if demand studies of outdoor recreation should be considered for publication in its principal journal or for professional awards. Yet by the late 1970s and early 1980s, the Association of Environmental and Resource Economists (AERE) had come into existence, with its own scholarly journal and awards.

We were fortunate at OSU that our graduate program in resource economics was begun in the 1950s, just as the paradigm shift got underway. It was also fortunate, I believe, that the program did not require that students conform to an arbitrary mainstream. By the 1970s, when the second paradigm shift was occurring, several graduate students from our program were prepared to assume leadership in the profession (Nelson, Norton, Norman, Stevens, Sokoloski, Bromley, Batie, Beattie, Randall, Ervin, Byerlee, among others). Knowingly or not, they contributed to the paradigm shift that occurred.

UNIVERSITIES IN SOCIETY: Three assignments during my first 21 years at Oregon State taught me much about universities—my service as dean of faculty, the Goals Commission experience, and my time as graduate dean. I recount some lessons learned and describe some examples to demonstrate that academic administration can be intellectually challenging. I have drawn heavily on the philosophy of science and economic theory. Yet there are many fields that provide powerful and useful administrative insights. Unless administrators can find intellectual excitement in the work they do, perhaps they should defer to someone who can. One of the saddest situations I have observed in my entire academic experience involved the provost of one of the largest and more prestigious institutions in the land. He was a most accomplished and eminent scholar in his chosen field of chemistry. He said to me in a one-on-one conversation, "I find no intellectual stimulation in anything I do."

When I was dean of faculty, I was expected to be of assistance to troubled academic units. These troubles were sufficiently serious to require help from the highest administrative level in the university. It is not surprising that, as an economist, I examined economic theory for

possible help. I was never so naïve as to believe there might be something like an optimal size of academic unit. (I do not think any such
thing exists even with market-oriented firms, except in the purest of
theory.) Nevertheless, my theory did lead me to speculate as to whether
there was any relationship between the size of the academic unit and
the difficulties that academic units might experience. Such difficulties
were of varying nature, including internal squabbling, inability to stay
within budgets, and student-faculty conflicts.

My limited sample suggested that troubled academic units tended to
be disproportionally small or large. At that point, my theory was helpful in identifying causes. In small departments, a few individuals often
can dominate policy for the entire department. This is not necessarily
bad if the dominant group is benevolent, but that will not necessarily be
so. Personal animosities in small departments can become exceedingly
intense. I had one problem department with nine faculty. Three opinion
centers emerged, with three faculty in each group. Enormous bitter-
ness and anger were associated with every group decision they tried
to make, and they were quite literally unable to make any important
decisions. Large departments may have severe problems as well. Here,
inadequate communication often exists that manifests itself in actual or
perceived feelings of isolation.

As noted earlier, I was one of three campus-wide academic administrators when I was dean of faculty. A current administrative chart for
Oregon State, or any comparable institution, provides an unbelievable
contrast. I do not imply there are too many people doing academic
administration. I do not know if that is true or not. My concern is different. I have observed numerous administrators who believe their principal responsibility is to review, and possibly modify, the decisions of
those who report to them. That multilevel activity can get out of hand
as additional levels of administration are added. When I was department head, I expected my decisions to be reviewed by the deans and
directors to whom I reported. But during the six years I was head, I was
unable to get all of those to whom I reported in the same room at the
same time for agreement on the long-run objectives of the department
I headed. My perhaps impractical suggestion for improving academic
administration is to require every newly appointed administrator to

write an essay addressing the question: "What can I accomplish in this position that is not likely to be addressed by any other administrator in this university?" Every administrator to whom the new administrator reports would be required to write a critical review of the essay for each administrator's personnel file before the new appointment can take effect.

The United States is blessed with great diversity of colleges and universities. I am always surprised and impressed by the differences in the places where degrees have been obtained by leaders in various lines of endeavor. This seems to be consistent with the belief that the United States is a land of opportunity and that there are many avenues by which success can be obtained. There are excellent four-year liberal arts institutions, and community colleges where the education is tailored to the needs of employers in the surrounding area. We have both private and public institutions. Two-year attendance at many places will earn a certificate of accomplishment in some subjects. At the other extreme, an enormous range of advanced degrees is offered by the system. There is much to criticize if an elitist view is taken. Yet, viewed from the standpoint of those who wish to use education to better themselves in the economy or in society, such an abundance of choice is a source of strength.

Despite this diversity, when viewed from an external vantage point there are motivations from within the system for greater uniformity. This motivation stems from what is sometimes called the academic pecking order. It cannot be denied that certain institutions have accomplished great things by their continuing focus on excellence. Harvard, Stanford, the University of Chicago, the University of California at Berkeley, the University of Michigan, and other comparable institutions provide outstanding service and deserve admiration. There is an understandable tendency in much of academia to emulate these institutions in great detail. Such a tendency must be carefully managed lest it become destructive rather than constructive. Excellence as an ideal need not be sacrificed if it is applied in equal measure to various activities. In other words, Oregon State need not have the same mix of activities as Stanford in order to conduct its activities with a comparable commitment to excellence. If this is recognized, the accomplishments at the elite institutions can be built upon in related, but not necessarily identical, endeavors. The implementation of such an approach means hard work for faculty, students, and administrators. Decisions must be made as to where and how excellence is to be achieved. Incentives and rewards should be consistent with those decisions. The output from, and individuals at, the more elite institutions are likely to be relevant and useful. This is so because they are excellent but different, not because of a motivation to be identical.

GRADUATE EDUCATION: One of my most satisfying activities at Oregon State has been the opportunity to work with outstanding graduate students from around the world. These opportunities arose in connection with graduate education in my department and in my time as graduate dean. My readings in the philosophy of science have been useful as I have retrospectively evaluated those experiences. Kuhn's work has been especially helpful, especially the distinction he makes between revolutionary and normal science as well as his "disciplinary matrix." The academic environment in higher education generally is now much more accommodating of graduate students than when I came to Oregon State. And conditions at Oregon State when I came in 1954 were better than they were at numerous other institutions at that time. This was due in no small measure to the leadership of my predecessor as graduate dean, Henry Hansen.

Perhaps the most important generalization I developed from my experience with graduate education is that it involves the acquisition of a culture and requires socialization as well as disciplinary competence. From this generalization, important deductions can be developed:

- Graduate education should insure that the student has a perspective from which to place the discipline or field in a broader context. There are various dimensions from which a broader context may come:
  - *Time*. History uses time as context. For example, Chapter 3 of this book presents a brief description of the English-American tradition in economic thought from Adam Smith to the present.

If such information were combined with economic history of the period in which these economists lived and worked, contemporary economic thought could be placed in a broader context than can be provided if the passage of time is ignored.

▢ *Place and circumstance.* Intellectual efforts are affected by the environment in which such efforts occur. For example, how were economic problems viewed in places and circumstances other than in England and the United States?

▢ *Concept.* When I was graduate dean, I asked a physical science department head if graduate students in his department were familiar with philosophy of science concepts. His reply: "Oh yes. We acquire such knowledge from our water cooler conversations." Even advanced graduate students often do not appreciate how the conceptual base of their discipline or field influences research results and what is considered important. A simple demonstration will illustrate the point. Consider, say, three acres of undeveloped land that is being considered as the site of a municipal building. Then ask workers in the following fields and disciplines to write a paragraph describing what they see as they observe the site: geography, botany, urban planning, entomology, political science, ecology, and regional science. A major contribution of the philosophy of science is to provide a systematic way of placing one's own field or discipline in an encompassing context.

• Create a learning environment in which a decent respect for all academic fields and disciplines exists or may be acquired. The view graduate students have of other fields is often influenced greatly by the attitudes of major professors and other faculty. Unfortunately, some believe that respect for one's own field or discipline can be acquired only if others are disparaged. Differences in subject matter require differences in the way knowledge is acquired. Every field and discipline has its limits, but the way such limits are revealed varies from field to field and from discipline to discipline. There are numerous ways in which graduate students learn about how their specialties compare with other bodies of knowledge. I believe graduate students learn much from one another. An important part of graduate education may involve bringing students together from different fields and disciplines. They can be informed that they are in the process of being assimilated into a community of scholars that has a broader base than exists within their own field or discipline.

- Encourage a learning environment that discourages and penalizes the use of graduate students as pawns in faculty disputes. This deduction is so obvious that I am somewhat embarrassed by including it. Yet I have observed great damage when it has been ignored. Graduate students understandably wish to satisfy major professors and faculty with whom they work closely. They may do the bidding of faculty even though they may not understand why they have been asked to do so, or they may not even approve of what they have been asked to do. In my view, such manipulation of graduate students is unprofessional in the extreme. Administrators can forestall such practices by stating their unprofessional nature clearly and unequivocally. It is well within the province of academic administration to do so.

5

RESOURCES FOR THE FUTURE
(1975–1986)

In late summer 1975, I received a telephone call from Marion Clawson asking if I would interview for the position of vice president of Resources for the Future (RFF). Marion was a well-known agricultural and resource economist and at that time was interim vice president at RFF, after having served as RFF's interim president. While RFF is known in those academic and policy circles concerned with natural resource and environmental issues, most people have never heard of the organization. If the connection of RFF to my intellectual journey is to be understood, it is necessary to know how RFF came into existence, the way it developed, and its characteristics in 1975.

WHAT IS RESOURCES FOR THE FUTURE?

During and after World War II, there was concern in the United States that the U.S. economy might revert to the stagnation and unemployment that existed prior to the war. As we now know, rapid growth and inflation instead became prevalent. Concern then began to emerge about natural resources and materials constraints that might arise. In 1950, President Harry Truman established a Materials Policy Commission, chaired by William Paley, Columbia Broadcasting System CEO, to investigate the long-range outlook for materials and supplies. In order to conduct the necessary investigations, a staff was assembled of well-

Theodore Schultz   Gilbert White

established scholars and others who were later to achieve prominence. The Commission issued a report early in 1952 under the title *Resources for Freedom.*

At about the same time, the Ford Foundation emerged as a major player on the national and international scene. It was first established in 1936 with a relatively small gift from Henry Ford. During the 1940s, other gifts from the Ford family combined to make it the largest philanthropic organization in the world. By 1950, the Ford Foundation was finalizing a formal program for advancing human welfare. Conversations developed between Ford Foundation people and natural resource leaders about possible programs that would build on the work of the Materials Policy Commission. A group of 26 people subsequently approached the Ford Foundation with a proposal to establish a nongovernmental organization to monitor the nation's natural resources on a continuing basis. The new organization came into existence in October 1952 and was given the name Resources for the Future. It immediately became active and helped plan and execute a midcentury White House conference on natural resources early in President Dwight Eisenhower's administration.

Milton Eisenhower, William Paley, Theodore Schultz, and Gilbert White were among the 26 people to approach the Ford Foundation.

Milton Eisenhower was then president of Pennsylvania State University, and William Paley became the first chair of the RFF board of directors. Both Theodore Schultz and Gilbert White were later to affect my intellectual journey in direct and important ways.

The Ford Foundation not only founded RFF but maintained it financially for more than 25 years with a series of four-year grants. By 1975, when I received Marion Clawson's telephone call, RFF had published approximately 175 titles pertaining to natural resources and the environment. Every graduate program in academia concerned with natural resource and environmental policy made use of RFF publications.

Rueben Gustafson, president of the University of Nebraska when RFF came into existence, became RFF's first president. Joseph Fischer was vice president and became president when Gustafson retired four years later. Fisher continued as president until 1973.

RFF governance developed in an unusual way. There has been an RFF board of directors composed of notable people throughout RFF's existence. Yet the real power rested with the Ford Foundation, because it provided the funding. Negotiations often occurred between the RFF president, the RFF board chair, and a Ford Foundation officer with responsibility for RFF supervision. As nearly as I was able to determine, the Ford Foundation greatly influenced final decisions on most important policy matters. At the time I went to RFF, and for some years earlier, Marshall Robinson, Ford Foundation vice president, had responsibility for RFF. When Joe Fisher left RFF in 1973, there were rumors that the Ford Foundation desired change at RFF. Those rumors suggested the Ford Foundation wished for RFF to become less academic and more policy relevant. There was considerable delay in naming a president after Fisher resigned in 1973. Marion Clawson was interim president between Fisher and Charles Hitch, who did not occupy that position until July 1975. Hitch had been RFF president approximately two months when Marion Clawson called me about interviewing for vice president.

As nearly as I was able to determine in my conversations with Marion Clawson, the issue of the Ford Foundation's wish for RFF to become more "policy relevant" had largely been ironed out when Charles Hitch agreed to become president. I later learned that was not the case. I also learned that this was far from a simple matter, and that well-meaning people often would "talk past one another" when the issue was being discussed. Some background at this juncture will be helpful in appreciating the complexities involved.

Much early RFF work followed directly from the efforts of the Materials Policy Commission and addressed such questions as: "Is the United States likely to exhaust its stock of iron ore, copper, or farmland if post–World War II economic growth continues at a rapid rate?" To address such questions even on a superficial level, a great deal of historical data had to be collected and connected to assumptions about future use and needs. Even when such data were collected and analyzed, numerous questions arose: What about the development of substitutes associated with different technologies? How will future economic growth affect the composition or combination of resources that will be needed or desired? RFF President Gustafson, Vice President Fisher, and others responsible for RFF programming recognized that these were complex issues requiring fundamental intellectual effort in more than one discipline.

The early RFF staff members were given freedom to open up new areas of investigation. Their findings often directed attention to the need for new public policies, as well as to the inadequacies in existing policies. Energy, minerals, land use, water resource development, outdoor recreation, and urban development were subjects for investigation. Much of this early work was highly visible in policy circles because it was new, it was interesting, and it was of obvious importance. The effort was generally of high quality, and it attracted attention because it was applied to previously neglected problems. In the beginning, RFF involved several disciplines, although economists have always played a leadership role. As the organization matured, economists and economics became increasingly dominant.

These early RFF pioneers and scattered scholars across the land and in other countries began to recruit younger people to these interesting issues. Understandably, the subsequent investigations became more technical and complex. Because the research was often of a pioneering nature, existing outlets for disseminating the results were inadequate. RFF developed a relationship with the Johns Hopkins University Press for the publication of RFF books in both hard and soft cover.

The Resources for the Future building with courtyard   Marion Clawson

Established academic outlets of printed material on such subjects eventually became more accepting, and new outlets emerged. It became increasingly difficult in some instances to serve both an academic and a policy-relevant audience with the same publications. In the beginning, RFF did not need to be excessively concerned with this distinction. As the organization matured, it was essential that it become concerned.

Many, if not most, researchers and scholars like to address and solve interesting technical problems within their fields. When doing so, they may forget about, or neglect, the problem arising in society that motivated their investigation in the first place. When I first considered an appointment at RFF, I was well aware of these issues, and I could understand how they complicated the Ford Foundation relationship. Nevertheless, it was naïve of me to think they could be easily "ironed out." I will write more about this issue and the RFF research agenda later in this book.

WHO WAS CHARLES HITCH?

I had not yet met Charles Hitch in 1975 when Marion Clawson called, but I certainly knew a great deal about him. I first encountered his name when, as a graduate student, I read a journal article he co-authored with another economist (Hall and Hitch 1939). He obtained his education at the University of Arizona and Harvard before going to Oxford as a Rhodes Scholar. He was the first American student at Oxford to remain there as a don. When he returned to the United States, it was to head an interdisciplinary program at the Rand Corporation concerned with national defense. When John Kennedy became President of the United States, Robert McNamara became his Secretary of Defense. McNamara brought a group into the Department of Defense, headed by Charles Hitch, who became known as the "Whiz Kids." This group pioneered the use of cost-effectiveness analysis in Department of Defense decisions. Hitch later went to the University of California as comptroller, then succeeded Clark Kerr as president of the University of California. His tenure in that position coincided nearly exactly with Ronald Reagan's time as governor of California.

When Hitch became president of RFF, he declared his intention that RFF should become larger, more policy relevant, and more multidisciplinary. At the same time, the Ford Foundation announced that it had made an additional four-year grant in support of Hitch's announcement.

I knew of RFF's outstanding reputation in academia, was greatly impressed with Hitch's record, and knew of the Ford Foundation's renewal of financial support. Hitch's reputation and stature surpassed that of anyone with whom I had previously associated. It was not difficult to agree to an interview. At that time, I did not know how many candidates there were for the position, nor what my decision might be if an offer were extended.

AN INTERVIEW AND A DECISION

The possibility of a move to the Washington, D.C., area created mixed emotions in both me and my wife. On the positive side, the prospect of some years in or near the nation's capital would be exciting for both of us. Further, we both recognized that this would be a major career move. RFF's prestige and Charlie Hitch's stature were attractive attributes. Yet there were negatives as well. My wife had no siblings, and her aged mother lived in Corvallis. Our only daughter, Cheryl, and her husband were in nearby Portland. I enjoyed being graduate dean, but it was not clear the position would continue to be a challenge. I did not aspire to be a university president, but an attractive alternative to being dean was to return to my department as professor. The most pleasant, comfortable alternative was to remain at Oregon State University.

Merab accompanied me to Washington, D.C., for the interview, which went well, in part because I was familiar with the RFF history and the organization as it then existed. I obtained my doctorate in 1952, the same year RFF came into existence. RFF was becoming well known when I began a water economics research program at OSU in 1954. Subsequently, I came to know several RFF staff. Allen Kneese provided advice when our early work on water quality was getting underway. John Krutilla came to OSU as a faculty member for our 1969 summer session in natural resource economics. I also knew Marion Clawson and Blair Bower, RFF staff members. Gilbert White—internationally known geographer, then at the University of Colorado, and RFF board chair in 1975—and I knew one another as well. It is probable more than one of these people recommended me to Hitch as a potential RFF vice president.

An offer was extended near the end of my two days at RFF. After returning to Corvallis and reflecting on the offer, two concerns arose. First, my responsibilities as RFF vice president had not been made as clear as I would have liked. Second, I did not know, in depth, how the senior RFF staff would view my appointment. Additional negotiation clarified both issues. My wife and I discussed the decision at length prior to acceptance. We finally agreed we wanted the adventure of living in Washington, D.C., for a period. At that time I was fifty-two years old. We mutually agreed we did not want to be away from the Pacific Northwest forever, probably not more than a decade. I accepted the position and became vice president and senior fellow, Resources for the Future, on December 1, 1975.

Clearly, this was a significant change for me. Even though I knew RFF as an organization and had interacted with members of the staff, RFF and OSU were very different institutions. My early education was in a one-room country school. I was the first member of my family to acquire a four-year college education. RFF's origin lay largely in a different tradition. Many notable people had contributed to the RFF reputation over the years. Former members of the RFF board of directors included William S. Paley, Laurence Rockefeller, Robert Anderson, and Edward S. Mason. Mason also had been a member of the Materials Policy Commission.

During my 10 years at RFF, I came to know several members of

the elite Eastern establishment, but none impressed me more than Ed
Mason. When I went to RFF, he was an emeritus member of the RFF
board. He was educated at the University of Kansas, Harvard, and
Oxford. As an economist, Ed Mason founded the field of industrial
organization. He had worked in government and been a university
administrator and faculty member at Harvard. There was no evidence
he cared at all about my background or education, but he frequently
tested the quality of my ideas as he did those of others with whom
he associated. Gilbert White explained to me that emeriti RFF board
members were eligible to attend meetings but did not debate issues. He
said Ed Mason did not speak in board meetings except on important
issues, but when he did speak, no one had ever challenged his right to
do so. John Kenneth Galbraith is reported to have said, "Wherever Ed
Mason sits is the head of the table." I turned to him for advice many
times during my time at RFF. I am deeply indebted.

THINGS WERE DIFFERENT THAN THEY APPEARED

When I went to RFF full-time on January 1, 1976, the organization had
grown by more than 20 percent in the 6 months since Hitch became
president. The organization had originally been established as a multidisciplinary group. By 1976, economics had become the dominant
discipline. In the beginning there were active outreach programs into
academia to encourage natural resource–related research and education.
Some outreach programs still existed in 1975–76, but most effort was of
an in-house nature.

When Hitch became RFF president, formal program units did
not exist. Instead, there were informal subject-matter groupings that
changed from time to time. I once asked what process was used to identify new areas of work. The response I received was "cell division."
When I arrived, three divisions had been established, and an elaborate
personnel classification system was being implemented. Soon after my
arrival at RFF, I discovered that communication between Hitch and
many RFF staff left a great deal to be desired. Joe Fisher had had an
informal administrative style and little use for formal administrative
structure. Charlie Hitch was a product of more formal and much larger
organizations—Oxford University, the U.S. Department of Defense,

the University of California. I told Hitch that he should have more conversations, individually and collectively, with organization personnel. He replied: "Why? I talk to you every day, and you speak for me. You talk with the staff, and they talk with you." End of conversation.

This was my first surprise after going to RFF. Yet I never considered this an insurmountable problem. I understood what needed to be done to address the situation and believed I had the capacity to do so.

Charlie Hitch and I established a good working relationship. He was a better writer than I am and could make prose "sing." His was an excellent intellect, honed by the best education available. We became complements, perhaps because I am able to analyze complicated "messy" matters, extract essentials, and imagine possible alternatives. Hitch liked to have "talking papers" before interacting with others about important subjects. I attempted to anticipate emerging issues, identify ramifications, and isolate RFF options. Almost always I gave him my recommendations, but they were separate from the analysis I had prepared. He usually would take my recommended course of action and, after discussion, build upon it in an elegant and substantive way. I sensed he was pleased with my work, although he seldom evaluated my performance. Once, almost inaudibly, he muttered something like: "You are as good, or better, than any associate I have ever had."

My wife and I became very fond of Charlie and Nancy Hitch. He was never popular with some of the staff, but those people were willing to deal with me. I soon learned Hitch wanted to convey good news, but that I should be the bearer of most sad tidings. I did not mind this arrangement, but I believe such a policy is a mistaken one. A leader is most respected when he, or she, announces all major developments to those for whom they have responsibility.

Approximately three and one-half months after I went to RFF, Marshall Robinson, vice president of the Ford Foundation, came to RFF to meet with Charlie Hitch and me. Robinson told us that McGeorge Bundy, Ford Foundation president, was making plans to retire, probably in 1979.[2] Robinson said it was unlikely, after Bundy retired, that the Ford Foundation would provide the unrestricted financial support RFF had enjoyed for the past 25 years. Robinson proposed a course of action to help RFF establish an endowment and seek program funds from other sources. He said RFF should immediately become smaller and undertake an active fund-raising campaign.

Hitch told Robinson this made his own position a very different one than had been agreed when he accepted the offer to become president. Robinson did not disagree but said these were the realities. After Robinson left, Hitch told me he had no inkling anything like this was in the works when we were negotiating the RFF offer for me to become vice president. Clearly he was concerned about my welfare, and I was touched. I replied that I did not hold him responsible, and that realistic people knew we lived in an uncertain world. I sensed he was grateful for my attitude.[3]

THE STRUGGLE FOR SURVIVAL

The Robinson message changed greatly the leadership needs at RFF. Although this was not discussed openly, I knew this to be the case immediately. The organization had enjoyed unrestricted funding for 25 years. RFF had a competent staff with an excellent reputation, but there was little experience among the staff and no infrastructure within the organization for seeking funds. Further, the RFF board of directors did not have a tradition of concern about the survival of the organization because of the Ford Foundation's role as patron saint. There was general agreement that our credibility was our most important asset, an asset we needed to preserve. This meant RFF must retain its independence in deciding what investigations it would undertake, have freedom to publish in the public domain, and maintain an arm's length relationship with those providing funding. When I was at RFF, staff consulting was minimal and tightly controlled; officers and staff were not permitted to work for, or take fees from, any organization or person that might be affected, positively or negatively, by RFF research. During my tenure at RFF, I dealt with financial supporters in government, private industry, and public foundations. Attempts were made from within each of these groups to influence RFF research results or how we would conduct research. I believe none was successful.

To his credit, Charlie Hitch promptly informed the staff of the message Marshall Robinson had given us. He and I agreed that reduction in size was necessary, and I began work with the three division directors to make program adjustments. Endowment fund-raising did not go well. We hoped the "big three" former RFF board members— William Paley, Laurence Rockefeller, and Robert Anderson—would become endowment sources. Each helped, but not enough to establish a significant endowment. We had some success in obtaining project and program funds, and the Ford Foundation made a retrenchment grant that established a reserve fund and served as an endowment base.

The challenges I faced following the Robinson visit were altogether different from those I had confronted as an OSU department head and dean. Instead of the excitement associated with building new programs with young people, I needed to retrench a mature, established staff. Not surprisingly, morale took a nose dive. I found myself not only holding hands with young staff, but also reassuring some who had created classic literature. Bad news travels fast, and the best people at RFF began getting "feelers" from prospective employers.

Charlie Hitch approved a plan of action I developed. The division directors and I estimated the number of core research leaders, or critical mass, we believed necessary for RFF to remain relevant in the public policy arena. Our estimate fell in the 15 to 20 range and, as I recall, the specific number selected was 17. I then estimated the number of years the core group, together with needed support staff, could be sustained by our reserve funds and realistic "soft money" expectations. This turned out to be eight to ten years. The next step was to identify the 17 people to be protected. I selected and then conferred with three. The four of us conferred and reached agreement about another group of five. We continued in this way until 17 had been selected. These core group members were assured of employment at RFF for eight years. After this core group was made secure, we explained the plan to all employees. Employees not in the core group were assisted if they wished to seek other employment. Several did well raising funds in support of their programs and remained at RFF. Morale improved immediately when the plan was announced and actions were taken to implement it. Administrative policies can create, reduce, and affect the incidence of uncertainty. Transparency, or lack thereof, also affects how employees perceive and interpret uncertainty.

From 1976 through 1978, RFF struggled to adapt to its new

environment. The plan described previously was designed to get the organization through a rough period, but it was not necessarily sustainable. Charlie Hitch and Marshall Robinson devised a different approach and proposed a possible merger with the Brookings Institution. The RFF reserve fund, together with a $7 million Ford Foundation grant, was to go to Brookings as a dowry if they would accept RFF as a distinct division within its organization. Brookings reacted positively to the proposal. The next step for Hitch and Robinson was to obtain reactions to the proposed merger from the RFF board and staff.

Early in 1978, Charlie Hitch told Gilbert White, RFF board chair, and me about the proposed Brookings merger just before an RFF board meeting. Because it came as a complete surprise, neither the RFF board, nor the RFF staff, was prepared to react immediately. During the ensuing RFF board meeting and staff discussions, it became apparent that the board, staff, Gilbert, and I were not favorably disposed to the merger. I was frank with Charlie Hitch about my concerns, and this difference of opinion affected our relationship for the next several months. Despite my concerns, during the summer of 1978 I worked closely with the Brookings Institution on the program and administrative details of a possible merger. As the plan developed, it provided I would lead the new Brookings division with the title of vice president, the only vice president Brookings would have. In midsummer 1978, Hitch told the RFF board of directors of his intention to retire in the summer of 1979. At the October 1978 RFF board meeting, I was named president designate, to become president when Hitch retired.

In the summer of 1978, the RFF board asked the Ford Foundation what it would do for RFF if we were to cut all apron strings and become entirely independent. We hoped they would grant us the $7 million they had offered as a dowry if the merger were consummated. Instead, the Foundation decreed the $7 million could become a matching grant if RFF matched it within 90 days. Of course, this was highly unrealistic, a fact well known at the foundation. Nevertheless, by August 1978 Gilbert White had received a $2.5 million commitment from another major foundation. He was not at liberty to reveal the source, and initially chose not to announce the commitment. He and I discussed how this information might be used to persuade McGeorge Bundy to extend the matching time limit from 90 days to a year or more. We knew Bundy, Marshall Robinson, and Charlie Hitch would be at a meeting at Aspen, Colorado, in late August. The problem was that Gilbert could not attend that meeting because of an international commitment. He said I must go in his place and request more time to meet the match. He said my presentation should include my plans for RFF as an independent organization. At that point in our discussion, something like the following exchange took place:

> EMERY: "Gilbert, do you expect me to go to a meeting with Bundy, whom I have never met, that also will include Robinson and Hitch, who favor the Brookings merger, to ask for an extension of time to meet a match they all hope will fail?"
> GILBERT: "Emery, if you expect to be president of RFF, you had better start acting like it."

The meeting with Bundy was in his cabin in late afternoon. Robinson opened the meeting and asked me to brief Bundy on merger plans with Brookings. When that discussion was completed, I asked if we could discuss an independent RFF. There was a moment of silence, but Bundy agreed, noting only that we would need to go to dinner soon. Bundy had discussed an independent RFF with RFF supporters, but he did not know about the $2.5 million pledge Gilbert had obtained. I told him of that grant and requested RFF be given a year to meet the remainder of the match. Bundy wanted to know where Gilbert had obtained the $2.5 million pledge. I told him I did not know, but I gave him a telephone number where Gilbert could be reached. I had deduced correctly from where the grant would come (the Andrew W. Mellon Foundation), and I sensed Bundy did as well. After considerable discussion, Bundy extended the matching-grant time period, but he stipulated October 1, 1979, was the absolute "drop dead date."

In addition to the $2.5 million pledge received from the Andrew W. Mellon Foundation, John Krutilla was instrumental in arranging a $1.5 million grant from the C. S. Mott Foundation in October 1978. On September 30, 1979, the last eligible day for matching, the John D. and Catherine T. McArthur Foundation responded positively to my written proposal for a $1 million grant. That grant, with other pledges, permitted RFF to meet the $7 million match.

LIFE IN AND AROUND THE NATION'S CAPITAL

Even though the RFF offer was accepted in 1975, I did not begin full-time work until January 1, 1976. Merab and I spent two weeks in the Washington area early in December 1975 and purchased a three-bed-room ranch-style house near McLean, Virginia, in the Salona Village subdivision. The subdivision had been established during the 1950s, when zoning rules were less well developed then than they were later. The houses apparently were located wherever the owners wanted their houses placed. When we moved there a quarter of a century later, the landscaping was mature, and most of the original trees, mainly oaks and dogwoods, had been preserved. Our neighborhood was beautiful indeed, especially during the months of April and October. As birds migrated south in the fall, large numbers favored us with their presence. The reverse occurred again in the spring when white dogwoods and multicolored azaleas synchronized their blooms. In the fall, trees and shrubs turned beautiful shades of gold, orange, and red as humans and other species readied themselves for winter.

Salona Village is bounded on one side by Dolly Madison Avenue. Ours was a community of relatively modest two-, three-, and four-bedroom homes. On the other side of Dolly Madison, the homes are larger and much more luxurious. Ted Kennedy's house was to be found there, as was the home of his sister-in-law, Ethel Kennedy. Our home was approximately 10 miles from my office near Dupont Circle, in the beginning on Massachusetts Avenue and later at 1616 P Street NW. To go from my home to my office required crossing one of a number of bridges that spanned the Potomac River. Bridges sometimes became bottlenecks if traffic was heavy because of inclement weather or for other reasons.

We were fortunate to have an enduring friend who lived near McLean and who also worked in the District. When I was an OSU department head, we appointed Gary Seevers to an assistant professor position in our department soon after he obtained his doctorate at Michigan State University. He had been with us about three years when he was invited to become an agricultural economist with the President's Council of Economic Advisors. I facilitated his leave from OSU because it was a remarkable opportunity for someone so young. Obviously his work was impressive to the people there, because they wanted him to remain as leader of the staff economists. He agreed to do so and resigned from OSU. Shortly thereafter he was appointed to the council itself. He was the youngest person to be appointed to the council, and the only person to be appointed to the council from its own staff. Later he became a member of the Commodity Futures Trading Commission. Gary and I commuted together for a period when we both lived near McLean. We remain in touch to this day. His friendship was of great value as we became acclimated to the Washington, D.C., community.

Two institutions stand out in my memory as we integrated ourselves into the Washington community—the Cosmos Club and the Kennedy Center. Marion Clawson nominated me for membership in the Cosmos Club, and I was elected shortly thereafter. It is a prestigious club with a membership drawn heavily from the academic, scientific, and professional communities. The Cosmos Club building was once the residence of Sumner Wells, a former Under Secretary of State. When I became a member, women were not permitted to become members, although this subsequently changed, after I returned to Corvallis. I entertained business associates, family, and friends at the Cosmos Club. It is located on Massachusetts Avenue, only a short distance from RFF offices. We also spent time and treasure at the Kennedy Center. We entertained there as well. We could take one or more couples to dinner there prior to an event in one of the three entertainment centers. During the course of any year, several world-renowned entertainers would appear there. There were three things I missed that I could not take with me when we returned to Corvallis—WGMS, a radio station with wonderful classical music, the *Washington Post*, and the Kennedy Center.

Jimmy Carter was elected President in 1976 and served until 1981. He was succeeded by Ronald Reagan, who was elected in 1980 and served until 1989. I was at RFF from 1976 to 1986 and was able to observe these two administrations at close range from a natural resource and environmental perspective.

Energy availability was a major issue during the Carter years and led to the creation of the Energy Department. Charlie Hitch and James Schlesinger, the first Energy Secretary, were friends. This permitted RFF to play a role as the department was being organized. Later we produced two outstanding books on energy that were directed at specific policy issues. It was clear our work was known in policy circles at that time. I had known a staff member, Lynn Daft, in the domestic policy office in the White House prior to his appointment to that position. At his request, I prepared a short paper on water issues and public lands in the West and took it to the White House for discussion with him and Stuart Eisenstadt, President Carter's domestic policy advisor. While there he took me to lunch in the White House mess. I had been there previously as a guest of Gary Seevers. The White House mess is an institution in itself in the Washington, D.C., arena. It is a white-tablecloth place with many tables of varying sizes. The people in charge clearly rule it with great firmness, as surely they must. Reservations must be made and strictly observed. Orders are served promptly; it is not the place for long, leisurely luncheon conversations. I judge having eaten there becomes a badge of distinction in some circles. Proximity to power is the way score is kept by many in the nation's capital. When I was RFF president, I received numerous invitations to White House functions related to natural resources, energy, or the environment. I have tried to keep in mind Harry Truman's distinction between the importance of a person occupying a position and the importance of the position one occupies.

During the week of Ronald Reagan's inauguration, I was favored at RFF by a visit from Aaron Wildavsky, noted political scientist. He had supported Reagan's election and was in Washington, D.C., to participate in inaugural events. He said he came to "meet RFF's new president." He came right to the point. He said he was aware there was concern in environmental circles about two new appointees of the Reagan administration—James Watt, Secretary of the Interior, and Ann Burford, administrator of the Environmental Protection Agency. Wildavsky said he could understand the concerns but hoped sight would not be lost of the long and admirable record of many Republicans as friends of the natural environment. He predicted that, sooner or later, the Reagan administration would reflect the traditional pro-environment stance of earlier Republicans such as Theodore Roosevelt, the Rockefellers, Russell Train, and William Ruckelshaus. I was to recall the Wildavsky conversation in connection with an event that occurred a short time later.

The office of Malcom Wallop, senator from Wyoming, called me as the Reagan people were settling into their new roles. The person calling said Senator Wallop wished to have a review of natural resource agencies by the senate committee he headed. He wanted to obtain the cooperation of RFF, the Conservation Foundation, and the Nature Conservancy in conducting a part of the review. The plan was to have the officers of each organization ask questions of various agency heads at a one-day seminar to be held in a Senate hearing room. The schedule was already set. As RFF's president, I would ask the first questions of the first agency head, James Watt, the newly appointed Secretary of the Interior. Such an assignment was consistent with RFF's public service responsibility and, of course, I accepted. I prepared well because I suspected there would be press coverage. I designed questions to permit Secretary Watt to describe plans he wished to advance for the public lands under the jurisdiction of the Department of the Interior. The questions were not intended to be hostile, nor were they. There were, in fact, press in attendance, including several television cameras trained upon participants. In response to my first question, the Secretary erupted into a tirade that included such statements as: "All you RFF people do is sit around and think of ridiculous questions!" My first reaction was dismay. Here I was, a new RFF president, squabbling with the Secretary of the Interior. As the Secretary continued to speak, I knew I needed to remain calm. When he finished, I went to the next question on my list. After we went back and forth a few times unproductively, Senator Wallop called a different agency head and a different person to ask questions. Surprisingly, there was not much press coverage of this event. But then, I was not well known, and there was no real substance to our public exchange. The press did well to ignore the incident.

I was there for the remainder of that day, internally miserable. The meeting adjourned about 4:30 p.m., and I went to my office to take care of a few matters before going home. Sometime after 5 p.m., the telephone rang in my office. I answered, "Emery Castle speaking." The reply was, "This is Jim Watt. I am told I was out of line this morning. If so, I apologize." I replied that an apology to me was not necessary, but I was concerned that his remarks created a public impression that there were problems between my organization and his. I said that as far as I was concerned that was not the case, and that we were in business to serve the public interest, including service to the Department of the Interior. He seemed contrite. He invited me to his office a few times after that for small group meetings or for lunch.

Aaron Wildavsky's message to me was prophetic. James Watt and Ann Burford did depart from their respective offices during Reagan's first term. William Ruckelshaus, who had been the first EPA administrator, returned for a second tour and served from 1983 to 1985.

An account of my time in the nation's capital would be incomplete without reference to the Alfalfa Club. I had read about the Alfalfa Club before going to Washington, D.C., but had never imagined I might attend one of its famous dinners. The motivation for the Alfalfa Club comes from Washington insiders who exercise continuing influence regardless of the party or personnel in power at any given time. As far as I know, its sole significant activity is an annual dinner for which invitations are sent to anyone its members believe are important to them. This includes the President, Vice President, members of the president's Cabinet, members of Congress, and members of the Supreme Court. I was invited by Elwood Davis, who was RFF's attorney when I was there. He was partner in a law firm that had long had a presence inside the Washington Beltway. The annual black-tie dinners defy my descriptive capacity. I do not have specific information about the money or influence required to stage such events, but both must be enormous. The dinners begin about 8 p.m. with a dramatic appearance by one of the military service bands, employing drums, cymbals, and brass. A multiple-course dinner follows, with much time between courses for a few speeches and much table hopping. This time permits contacts and conversations outside the confines of government bureaucracy—perhaps the principal purpose of the dinners. The President, Vice President, Cabinet officers, Congressional leaders, and members of the Supreme Court are seated at an elevated table that runs the full width of the dining hall. Tables at a lower elevation are placed at right angles to the head table. These lower tables each accommodate upward of 30 people. When I attended, my table was directly below where President Ronald Reagan was seated. Between one of the dinner courses, my host directed my attention to a few people seated at other tables parallel to our own and asked if I knew them. They were people who had served as Cabinet members in previous administrations. After I said who they were, he then said, "Last year they were up there, where the President is seated. This year we have better seats than they do." His comment captures the essence of the Alfalfa Club. The dinners I attended concluded about 11 p.m. At that time, many of the participants went to the top floor of the hotel where the dinner was held. This, I was told, was where the serious deal making occurred. The one time I went to the upper floor, there were abundant refreshments, especially of a liquid nature. The alfalfa plant is noted for the long distances it will travel through the soil to obtain moisture.[4]

I was privileged to participate in a dinner of an altogether different nature held for members of Congress in the Woodrow Wilson Room in the Library of Congress. This is indeed a most beautiful room and was a wonderful setting for the event I experienced. I was one of a group asked to prepare discussion papers about various energy problems and policies. My topic was energy and agriculture. The paper went to the Library of Congress in advance of when I was to appear at the dinner for discussion with members of Congress, then, on the following morning, I met with Congressional staffers for further discussion. The dinner was surprisingly well attended by approximately 50 members of Congress. I sat between Senator Richard Lugar of Indiana and Representative Al Gore of Tennessee at a table for eight. The discussion was interesting, lively, and frank. I knew Senator Lugar at that time was acting as campaign manager for Howard Baker, who was seeking the Republican nomination for President. I asked him about Baker's prospects. His reply was, "Not good." I was impressed by Dick Lugar's frankness and wisdom. I also was impressed by how much Al Gore knew about my nephew, an intern in Gore's office as part of his Bucknell University program, and by Gore's obvious ambition. The evening was special, perhaps, because there were no reporters, and the members were not on guard with one another. There was warmth among them and genuine curiosity about what I might offer that would be useful to them.

AN INDEPENDENT RFF

The post–October 1, 1979, period was unique in the history of RFF. It was then fully independent except for some strings the Ford Foundation attached to its final matching grant. The foundation correctly noted that RFF had no experience managing endowment funds and stipulated that until all matching funds were received, exceedingly conservative investment guidelines must be observed. Even so, it was now clear that RFF was, for the first time, responsible for its own survival both financially and programmatically. We worked very hard to meet the $7 million match with real assets instead of pledges. We were able to do this by obtaining soft money rather than spending endowment income wherever possible. We managed carefully and controlled overhead expenditures in a variety of ways. Income from the endowment, as well as the value of the endowment, increased when we no longer were required to follow Ford Foundation investment guidelines.

In 1979, Gilbert White reached the age of seventy, the mandatory retirement age for RFF directors. This not only created a board vacancy, it also required a different board chair. He was succeeded by M. Gordon "Reds" Wolman. Reds was a member of the Johns Hopkins University faculty and an accomplished geographer and engineer. It would not have been possible for me to have had better board chairs than Gilbert and Reds. I am deeply indebted to both.

When Hitch left RFF in 1979, I assumed the duties of both president and vice president. Consistent with the responsibilities of these positions, I served as chair of the publications committee and made final decisions on the publication of all manuscripts. I also participated in considerable fund-raising. These were indeed busy months and years.

When I went to RFF, I did not believe that the administrative responsibilities of being vice president would occupy all of my time. I welcomed this possible freedom and anticipated I would establish a personal research program. When it turned out that I discharged the duties of both president and vice president, I acquired considerable subject-matter knowledge about the entire RFF research agenda. The research program was organized into three divisions: Renewable Resources, Environmental Quality, and Energy and Materials. In addition to my internal responsibilities, I spent time fund-raising as well as participating in various public-service activities. The latter provided opportunity to enhance the policy relevance of RFF work. As one example, while I was at RFF, the National Academy of Sciences undertook a major activity entitled the World Food and Nutrition Study, and I served as chair of the research priorities committee. RFF undertook and completed two major energy studies while I was at RFF. One was funded with a special grant from the Ford Foundation, the other by the Andrew W. Mellon Foundation. The latter was in the final stages and in rough-draft form when I developed a contact in the White House. This made it possible for me to get the draft into President Jimmy Carter's briefcase as he was leaving for a short vacation. Other RFF staff members also had contacts with members of the administration, so I would have been at a loss to identify the specific source of any impact, but the study was a good one and probably did have some influence on the administration's energy policy program.

During my time at RFF, I discovered that the emerging literature in economic research methodology could be a source of escape from my professional responsibilities. I never discovered any deep interest by anyone at RFF in this literature, but I knew Thomas Kuhn's work was having an impact on the philosophical base of economic research. Bruce Caldwell's *Beyond Positivism*, first published in 1982, stimulated me a great deal. Donald (later Deirdre) McCloskey's "Rhetoric of Economics" was published in the *Journal of Economic Literature* in 1983 and was to have considerable impact. Correspondence with a former student, Alan Randall, provided me with another important stimulus. He had taken my methodology course years earlier as a student at Oregon State and was making significant contributions to the field as faculty member at the University of Kentucky and then at Ohio State University.

Shortly after the matching grant was met in 1979, John Herbert, RFF treasurer, announced his retirement. His replacement was Edward "Ted" Hand. Ted was then in his early thirties. My recommendation to the RFF board for the appointment of Ted Hand as treasurer was the best decision I made for RFF. He added enormously to the capacity of the organization to meet the challenges that arose after he was appointed.

Ted and I conferred extensively on an investment strategy for the RFF endowment. I included real estate among the alternatives to be considered. RFF was then renting space from the Brookings Institution. It occurred to me that we might be able to acquire a building that would provide us with office space and supplement our income. I had developed great respect for Ted Hand, but until then I had no inkling how much he would add to RFF's capacity to acquire and manage financial resources. It turned out that he would leverage our endowment funds to acquire real estate, and then make that real-estate investment a major source of wealth and income for RFF.

The real-estate opportunity that arose was with a Truckers Association headquarters located nearby (between O and P Streets in one direction and 16th and 17th Streets in another, within the northwest quadrant of the District of Columbia). The building was on approximately an acre of land, mostly undeveloped. The National Wildlife Federation had its headquarters on adjoining land. As this opportunity ripened, the newly elected Reagan administration was becoming established. Monetary policies began to tighten, and fewer funds were available for real-estate development. When developers learned about the RFF balance sheet of solid assets and no significant liabilities, many were eager to help us acquire real estate. It turned out the National Wildlife Federation wished to partner with us if we were to acquire the Truckers building and develop the land. Its existing building was inadequate. The plan that emerged was to raze that building and build a new office structure in its place. Some of our undeveloped land was used for a multistoried residential structure. Underground parking was required. The renovation of the Truckers building interior would complete the project. When renovated, the Truckers building would provide RFF offices and office rental space as well.

This was an ambitious undertaking, of course, and it was not clear how either the board or the staff would react. Ted and I talked at length about the best procedure to follow. We had been transparent about our consideration of the project with both our staff and the board. Ted and I analyzed the possible purchase and development independently. His analysis emphasized most probable outcomes. I concerned myself mainly with the risk the real-estate investment would have for the endowment. We independently concluded that the opportunity, including staff reaction to it, should be placed before our board.

In preparation for the staff meeting, I recommended to Ted that he have his numbers on probable outcomes readily available. This was a staff that had great expertise in benefit-cost analysis and project evaluation. Some staff were supportive, while others were willing to defer to Ted and me. There was a vocal minority group of research economists who opposed the investment. They had little interest in Ted's numbers. Instead, they argued the investment could not possibly be a good one because, if it were, efficient markets insured that others would have seized the opportunity. Paul Portney, later to become RFF vice president and president, was a vocal leader of this point of view. The RFF board accepted Ted's and my recommendations and authorized the project. It has proved to be a great RFF investment.

The Marshall Robinson visit in 1976 changed my responsibilities from what had been mutually agreed to a few months earlier. My first responsibility (unstated) became the survival of RFF, rather than that of program officer of an established organization. By March 1986, RFF had met all of its pledged obligations to Ford, had developed real estate with a renovated office building in a prime location in urban Washington, D. C., and had assets of approximately $30 million. RFF had become a fully independent, endowed organization with its reputation intact. Marion Clawson said two people had been essential to RFF's survival—Gilbert White and me. If that was correct, my primary mission had been achieved.

I had begun to think seriously about leaving RFF early in 1985. Merab's health was in obvious decline. I believed it would be desirable to live in, or near, the place where we would spend the rest of our days (see appendix A). Later in 1985 I announced my resignation from RFF and left March 1, 1986, ten years and two months after taking up duties there as vice president.[5] That was only two months more than we had set as our goal for residence there when we decided to leave Corvallis. It was a great experience, but neither of us ever became infected with Potomac fever.

REFLECTIONS

My decade at RFF was extremely valuable. It made possible a comparison and evaluation of experiences in a land grant university with those of a small, prestigious "think tank." Conjectures that had arisen in one environment could be tested against developments in another. I learned a great deal about myself, our society, and the application of research and economic theory to practical affairs. Three such lessons stand out:

- The relation of theoretical models to real-world decisions.
- The close association of applied science to revolutionary or extraordinary science.
- Whether it is practical for an independent public policy think tank to pursue a nonadvocacy, nonpartisan research and education program.

A treatise could be written about each of these lessons. I introduce and comment briefly on each of them here.

THEORETICAL MODELS AND REAL-WORLD DECISIONS: Consider first the theory underlying the efficient-market model as a basis for actual individual decision making in contrast to its use for describing the performance of an economy over a long time period. This model depends on numerous assumptions, including the free flow of information and time for all parties to make adjustments to the information at their disposal. Efficient-market theory typically is separate from business-cycle theory, which, in turn, involves monetary and fiscal policy. As recent events have demonstrated, real-estate markets are indeed cyclical and much affected by monetary and fiscal policy. Markets are generally efficient, perhaps, but only in a long-run setting. Even then, perfect market theory rests on an inexact and incomplete theoretical base (Hausman 1992; Castle 2005). It is one thing to discuss such matters in the classroom or to use these ideas to design a research project. It is another to apply them in practice as, for example, by investing RFF's endowment funds in Washington, D.C., real estate. In recommending such a step to the RFF board, Ted Hand and I observed that few real estate developers had sufficient liquidity at that time to take advantage of attractive real-estate investments. In contrast, the RFF balance sheet was reasonably liquid and the organization was without debt. When economic models were modified by realistic conditions, both Ted Hand and I came to the conclusion that an urban real-estate investment in walking distance of the White House would not be a risky one. And, in fact, the investment has since proved to be a highly profitable one. Does this mean that if I were to work in urban real estate on a continuing basis I could "beat the market" or do better than those who work full-time in this arena? Of course not! Rather, the experience demonstrated that the "risky" use of general models occurs when they are applied in a mechanical way when theoretical assumptions deviate greatly from actual conditions.

Two additional RFF events deserve comment because they direct attention to the fragility of "best plans" in a dynamic environment. When I arrived at RFF in January 1976, the organization was busily making plans consistent with the arrival of two new officers and another four-year Ford Foundation grant. Within four months, those plans were shattered by the message brought by Marshall Robinson that McGeorge Bundy's retirement would be the end of the Ford Foundation largess for RFF. At that time, strategic planning was being advocated for all organizations, public and private. Would RFF have been well served if a well-developed strategic plan had been in existence at that time?

The Robinson visit resulted in a classic choice among short- and long-run possibilities. As noted earlier, it would not have been difficult to obtain short-run financial support by exploiting RFF's reputation of independence and lack of bias. Such a choice, of course, would have destroyed, or severely circumscribed, what RFF wished to maintain in a longer-run setting. Those doing strategic planning are not likely to envision that such a stark choice would arise. The Robinson visit required that those responsible for RFF's future make fundamental decisions as to the most important long-run characteristic RFF should attempt to preserve. The two words that occurred to me were "preserve credibility." As far as I was concerned, these words trumped any mission or vision statement I could imagine. They guided all important decisions I made during that difficult period.

It also was necessary that I address organizational issues when I became president. As noted, the RFF Board had not had responsibility for many traditional board functions, such as fund-raising and program policy. From the outset, I considered the board as our supreme policy body. The staff, by virtue of education and diligent effort, had responsibility for, and considerable autonomy in, the implementation of those policies. The officers were supposed to make the two come together. I believed it important to keep policy formation and its execution distinct. This worked well for me, perhaps in large part because of two superb board chairs—Gilbert White and Reds Wolman. Currently, there is discussion about collaborative efforts, at RFF and in other organizations, between research organization personnel and those with policy responsibilities in the planning and execution of research. The implications of this trend for policy research appear profound to me. As a consequence, I explore this subject in greater depth later, especially in Chapter 7.

Clearly, the organizational model I employed is not without its difficulties. When granted autonomy in the conduct of research, research personnel may fail to recognize certain fundamentals. The research they most want to do is not necessarily research that addresses the social problems that made funding of their research possible. This probably was at the heart of the reservations held by some Ford Foundation personnel about the RFF research program just before Charlie Hitch came to RFF.

APPLIED VERSUS EXTRAORDINARY SCIENCE: Consider next the close association of applied and extraordinary science. The RFF research program had a long tradition of investigating difficult or unexplored social problems associated with resources and the environment. When doing so, inadequacies were often encountered in the ready application of whatever underlying field of science was deemed most appropriate. To assist in such applications, outstanding scientists were often consulted and involved. In some instances, such people would assist with novel applications of existing theory. In other circumstances, attention would be directed to needed modifications or additions to existing doctrine. The RFF experience provided confirmation of what I had previously observed. That is, that the very best disciplinary scientists often welcome the opportunity to consult or collaborate in the investigation of problems in applied science. While at RFF, I came to know Nobel Laureates Kenneth Arrow and Tjalling Koopmans. And previously I had worked with Nobel Laureates Theodore Schultz and Leonid Hurwicz. These acclaimed economists were stimulated and complimented when confronted with an applied problem for which a ready-made theoretical solution did not exist. They knew our motivation was to make use of their imagination and intellectual power in addressing anomalies. I concluded that the tension I had observed between theoretical and applied economists at both RFF and land grant universities existed mainly among those of lesser accomplishment.

A COMMON UNDERSTANDING: ADMISSIBLE VERSUS INADMISSIBLE OUTCOMES: Even before I went to RFF, I was persuaded that society received benefit from the existence of an independent, nonpartisan, nonadvocacy natural-resource public-policy research institute. When its existence was threatened shortly after I went to RFF, my reaction to help it survive was largely instinctive rather than rigorously derived. Even so, that instinctive reaction was tested several times during the difficult days, weeks, and months that followed. I remain persuaded the survival of RFF was a desirable outcome.

The preceding paragraph does not stand alone; there is more to be said about this important subject. Even if a nonpartisan, nonadvocacy research organization exists, one cannot describe research results as "objective" except in the context of the viewpoint the researcher brings to the research. Irving Fox, RFF vice president for many years, has described the *implicit* premises he believed influenced the RFF research program for its first 25 years (Jarrett and Fox 1977, 69–73). These premises, or points of view, establish the legitimacy, or admissibility, of possible solutions to policy problems that are investigated. Conversely, other possible solutions become inadmissible. The implicit premises Fox refers to are those embedded in the efficiency model deduced from neoclassical economic theory.

Before going to RFF, I concluded that the viewpoint or premises of economists doing policy research often are, indeed, implicit. I concluded, further, that this was often the reason why the results of economic policy research became sources of controversy. The RFF experience validated, rather than modified, this conclusion.

If those doing, and those using, policy research do not have a common understanding of admissible or inadmissible outcomes prior to the conduct of research, the potential impact of such research will be diminished. The need for common understanding also exists among disciplinarians engaged in multidisciplinary policy research. No subject is of greater importance in graduate education, multidisciplinary cooperation, and public policy research. Only if there is agreement on what constitutes admissible and inadmissible outcomes has a basis been established for determining if research results are "objective." I elaborate further on this important subject in later chapters.

6

THE RETURN TO OREGON STATE UNIVERSITY
(1986–2010)

Merab's declining health made it clear that we now needed to be where we intended to retire (see Appendix A). I wished to continue to be active professionally and have time for support of Merab. She had not yet not been diagnosed with Alzheimer's disease, but it was obvious she needed more care than I was able to provide while serving as president of RFF.

My long-time friend, Carl Stoltenberg, was to play a key role in our return to Oregon State University. Carl, a noted forestry economist, had been dean of the College of Forestry at Oregon State since 1976. He was also a member of the RFF board of directors at the time. As a consequence, he was in a strategic position to facilitate our return to Corvallis and Oregon State under circumstances that suited our special needs.

Those responsible for graduate economics at Oregon State at that time were struggling. Three departments in three colleges were affected directly—Agricultural and Resource Economics in the College of Agricultural Sciences; Forest Resources, which housed forest econom-ics courses in the College of Forestry; and Economics in the College of Liberal Arts. Graduate degrees in economics were offered by both the College of Agricultural Sciences and the College of Forestry. Graduate degrees in economics had never been offered by the College of Liberal Arts because of the curricular allocation system described earlier. Economists in Liberal Arts participated in an interdisciplinary masters degree and taught courses for agriculture and forestry graduate students.

Several key faculty members in each of the three departments had recently left Oregon State. Economics graduate courses tended to reflect the interests of individual faculty members and were scattered among the three departments. These offerings, even taken in total, did not provide an integrated graduate student program. Deans of the three colleges identified above concluded that the resources of the University were inadequate to offer three uncoordinated graduate economics education programs. Their solution was to establish an interdepartmental committee to design a single core of graduate economics courses that could be taken by any economics graduate student on campus. The committee recommended that a part-time director be appointed and given the assignment to ensure that the newly designed course sequences be taught "regularly and well" in an Economics Core Program.

The three academic deans all had degrees in economics or economics-related fields. Bill Wilkins, dean of the College of Liberal Arts, was a PhD economist; Ludwig Eisgruber's terminal degree was in agricultural economics; and Carl Stoltenberg held a doctorate in forest economics. Dr. Charles Vars, professor of economics in the College of Liberal Arts, was chair of the committee appointed by the deans to design a core program. Dr. Richard Johnston, professor of agricultural economics, also served on the committee, as did Dr. Douglas Brodie, Department of Forest Resources.

The need for a part-time director of the Economics Core Program at Oregon State coincided with my anticipated departure from RFF. Carl Stoltenberg was aware of developments at both institutions and coordinated negotiations. Another long-time professional associate, John Byrne, had recently become president of Oregon State. He supported my returning to campus, as did the dean of the Graduate School, Lyle Calvin. I returned to Oregon State on March 1, 1986, with a half-time appointment.

Three professional assignments came my way after my return to OSU in 1986. They are described below in sufficient detail that lessons learned can be made explicit. They are presented in the following order: 1) The University Graduate Faculty of Economics, 2) the National Rural Studies Committee, and 3) Extended Education.

AN EXPERIMENT IN INTRA-DISCIPLINARY COOPERATION: THE UNIVERSITY GRADUATE FACULTY OF ECONOMICS (UGFE)

Two dimensions of the UGFE program are important here. One is the nature of the program; the other is its administrative structure. Each is discussed, but not independently. Because they are interconnected in reality, this is reflected in the following paragraphs.

Shortly after I became director of the newly constituted Economics Core Program at Oregon State in 1986, I took steps to enlarge its structure and focus. This was reflected in the new program name, the University Graduate Faculty of Economics (UGFE). This faculty group comprised those economists holding graduate faculty status in the departments of Agricultural and Resource Economics, Forest Resources, and Economics. The University administration moved promptly to approve this recommended change. One important reason for doing this was to provide for the growth of economics at Oregon State. For example, both the College of Business and the (then) College of Home Economics had expressed interest in later becoming a part of the University Graduate Faculty of Economics.

The three departments that initially constituted the UGFE could hardly have been more different. The Department of Agricultural and Resource Economics had the oldest and largest program. Graduate students in this department typically were much involved in research programs supported by both recurring and nonrecurring funds. This department had helped pioneer resource and environmental economics as a field within the economics discipline. The department also had an established reputation in traditional agricultural economics and at that time wished to strengthen its work in international agricultural trade.

One of the largest contingents of forest economists in academia nationally was to be found in the Department of Forest Resources at OSU. Not all of the graduate students in this department elected to work toward economics degrees. Some preferred to emphasize other areas of specialization, such as operations research or quantitative methodology, although some in these specializations still did considerable work in economics. The department also offered other degree options for graduate students, some having little to do with economics. Even so, this department provided long and dedicated support for the UGFE. The affected faculty believed strongly it was in their long-run interest for there to be a vigorous graduate economics program on campus.

The Department of Economics in the College of Liberal Arts did not offer graduate degrees in economics at that time but was a staunch supporter of the UGFE when I served as its chair. The UGFE, as originally constituted, permitted Department of Economics faculty to offer graduate-level courses and to serve on graduate committees. Numerous faculty in this department were engaged in productive research and scholarly activities. Such efforts were impressive in view of the large undergraduate teaching program for which the department had responsibility. When the UGFE was formed, many of the Department of Economics faculty were approaching retirement. Fortunately for me, many of these retirements did not occur until after I retired as chair. During the time I was chair, the department made an outstanding appointment in Joe Kerkvliet. The cooperation and support of the Department of Economics and the College of Liberal Arts for the UGFE was outstanding and never wavered during my tenure. This was exceedingly important for an applied economics program that depended on economics theory and principles to provide a strong conceptual base.

As the UGFE became operational, the Department of Economics requested permission to submit a proposal to the State System of Higher Education for authority to offer the masters degree in economics. Given that the State System had several times refused graduate degrees in the College of Liberal Arts at OSU, an alternative approach emerged. I visited quietly with a few key administrators and then consulted with the Department of Economics. I recommended that they request authority to offer masters and doctoral degrees in applied economics. I told them an applied economics graduate program would permit them to do research in three areas of specialization that had been established: 1) industry economics—forestry, agriculture, and fisheries; 2) resource and environmental economics; and 3) the economics of growth and change (economic development). I explained to them then, and more than once later, that these areas of specialization provided encouragement for economists on campus who were interested in industrial organization, the macroeconomics of housing and forestry, all aspects of resource and environmental economics, and traditional economic development, as well as the emerging new growth economics. After considerable discussion, the department agreed to prepare and submit a proposal to the Oregon State System of Higher Education for graduate degrees in applied economics. To receive approval, the program would need to be recommended by the chancellor's office and approved by the Oregon State Board of Higher Education. It was generally agreed that the position taken by the University of Oregon would be important. We hoped there would be no objection because we were seeking to offer degrees in applied economics, not a traditional economics degree.

When I became UGFE chair, I obtained external reviews of the core course work program that I had been employed to implement. I requested that the reviewers evaluate the learning that would be required of both masters and doctoral students in applied economics. The following provided such reviews: George Tolley, University of Chicago; Vernon Ruttan, University of Minnesota; Irma Adelman, University of California at Berkley; and Edwin Mills, Northwestern University. I also conferred by telephone with Kenneth Arrow, Stanford University, about the feasibility of a graduate program in applied economics in the profession. I informed the chancellor's curricular affairs office of these reviews. They never asked me about the review results, nor did they contact the reviewers themselves. The reviewers' comments to me were universally favorable, although there were suggestions for improvements in implementation, which I took seriously.

The University of Oregon did object to our offering advanced degrees in applied economics, but on a different basis than we anticipated. We had several meetings with U of O administrators and emphasized that we did not intend to offer degrees in traditional theoretical economics. They said that they offered "applied economics," not just economics. It turned out they were fearful that we would compete with them for applied economic research opportunities in Oregon and the Pacific Northwest. Finally, I offered to drop "applied" from the title of our proposed degree and seek an economics degree for our Department of Economics. To the best of my knowledge, the University of Oregon made no further objection to our proposal. I was astounded by the position they took. We did not rewrite our proposal, and as far as I know it remained applied economics throughout the approval process. I was not concerned about having offered to drop "applied" from the title of our proposal, because I knew the body of the proposal limited our emphasis to the applied fields we intended to develop at OSU.

The reaction of various parties to degrees in "applied economics," in contrast to "economics," is not easily explained. Some economists deny the wisdom of separating the two. Others would note that some economists have long argued that a distinction can and should be made (Backhouse and Biddle 2000). Attitudes on this matter change from time to time as new generations of economists reflect new perspectives. I was aware in 1986 that there were developments in academia providing greater respectability for graduate work in applied economics. At that time, the leading economics graduate programs were emphasizing mathematical economics theorizing. It began to appear that other programs could develop a comparative advantage with programs emphasizing real-world problem solving. I had concluded that Oregon State could never realistically aspire to be like Harvard, MIT, Stanford, or the University of Chicago, for example. Apparently, the University of Oregon had reached a similar conclusion and did not want the competition that might be provided by Oregon State.

There was some evidence that the academic affairs unit in the chancellor's office sought advice from the secretary-treasurer's office of the American Economic Association as to the academic standing of applied economics. Apparently they were told there was no such thing as applied economics. This failure to recognize a distinction has created enormous problems for economics graduate education at Oregon State.

Before making a recommendation to the State Board of Higher Education, the academic affairs unit in the chancellor's office sent us comments for review. The comments reflected a lack of understanding of how the UGFE was organized and what our objectives were. They assumed incorrectly that because, as UGFE chair, I reported administratively to the dean of the Graduate School, the UGFE degrees in applied economics would be awarded through the Graduate School rather than by the participating departments. In any case, the chancellor's office did recommend our revised proposal be approved, but in economics, not applied economics! When we received approval, to our astonishment all three departments—Economics, Agricultural and Resource Economics, and Forest Resources—were granted authority to offer masters and doctoral degrees in economics.

In the beginning, the UGFE decision structure was straightforward and transparent. Before agreeing to be director, I established the following requirements: 1) I would be administratively responsible to the dean of the Graduate School, 2) my office would not be in any one of the three departments constituting the UGFE, and 3) at least once each calendar year I would report progress and problems to the committee of four deans responsible for the UGFE—Graduate School, Agricultural Sciences, Forestry, and Liberal Arts.[1] The interdepartmental group that established the original Economics Core Program created the precedent that a majority of an interdepartmental policy committee would be required for decisions on curricular affairs. I chaired a policy committee that consisted of three members from each department in which most decisions were based on consensus. I do not recall taking any votes, but there may have been some.

During the six years I administered the UGFE, I drew great strength from the above described preconditions. Serving as the UGFE chair was not a time-consuming assignment. I was paid 25 percent of my annual salary to administer the program, and I believe that is about the amount of time I devoted to it. Yet this was as difficult an administrative assignment as I have ever had. I did not have a budget to use as leverage; budgets were with the academic deans and department heads. Faculty and department heads knew I had access to the deans, but I knew this card had to be played sparingly and only for important effect. I relied heavily on transparency and communication at all levels. I made sure that department heads and deans were aware of any potentially troublesome issues that I believed were arising.

We made progress during those six years. The Department of Agricultural and Resource Economics solved some internal problems and has made steady progress since then. The forest economics group made important faculty appointments. The Economics Department was made marginally stronger by the appointment of Kirkvliet and the presence of graduate students.

Soon after I departed as chair of the UGFE, tensions began to develop. The Economics Department faculty began attempts to establish a traditional economics doctoral degree within the UGFE. They wished to emphasize the doctorate rather than masters degree programs. This was consistent with traditional economics degrees across the land, where the masters degree is awarded as a consolation prize to unsuccessful doctoral candidates. In contrast, the other two UGFE departments had established markets in business and government for masters-level graduates in applied economics. The UGFE became increasingly bureaucratic with the passage of time. This was related to disagreements within the UGFE policy group. Increasingly, academic battles were waged there with written memoranda, written guidelines, and descriptions of requirements and areas of specialization. Differences of opinion within the Economics Department also emerged. Two full professors became marginalized. Starr McMullen requested a transfer from the Economics Department to the Department of Agricultural and Resource Economics. Joe Kerkvliet never returned to OSU after going on a leave, a major loss to the UGFE.

In 2005 it became apparent that the UGFE operation was becoming unglued. Opinions vary as to motivations and developments from that point on. Faculty in the Economics Department wished to add an advanced course in macroeconomics in the core program that would mainly serve their limited graduate enrollment. They maintained they did not have sufficient resources to teach core courses necessary for masters degree students. Their request was tabled in the spring for further discussion in the fall, in hopes of working out a compromise. Steven Buccola, professor of agricultural and resource economics, was UGFE chair at that time. He called on each of the four responsible deans and believed they were in support of the position that had been taken by a majority of the UGFE policy committee. The four deans met as a group and, without the presence of Buccola, reversed the position of the UGFE policy committee. Buccola resigned immediately as

UGFE chair. There were numerous attempts by the responsible deans to put the program back together. None were successful. Further, the deans were not of one mind as to appropriate remedial action.

The Department of Agricultural and Resource Economics and the Department of Forest Resources subsequently filed a proposal with the University Curriculum Council to establish doctoral and masters degrees at OSU in applied economics, supported by the College of Health and Human Sciences, which had expressed interest in the future development of health economics. This was matched by a proposal from the Economics Department for that department to offer doctoral and masters degrees in economics and econometrics. Obviously, OSU did not have sufficient resources to support two independent doctoral economics programs. The Curriculum Council asked the administration for guidance before it investigated the academic merit of either program. After this stalemate developed, the administration brought an outside group to campus to investigate. Its view of the matter was expressed in what has become known as the "Yardley Report," which endorsed the development of an OSU doctorate in applied economics, as was first recommended to the State Board of Higher Education in 1988.

Following receipt of the Yardley Report, OSU Provost Sabah Randhawa stipulated that an applied economics graduate program be developed consistent with a modified university applied economics graduate faculty framework. This has since happened and received final approval in February 2009, approximately three and one-half years after the UGFE became inoperative in 2005. The reconstituted Faculty of Applied Economics is more tightly structured than either the Core Program I was employed to administer or the UGFE that followed. The director of the new applied economics program will hold appointment in the Graduate School with his or her FTE supported by the Provost's Office. Most importantly, the director will have a budget that can be allocated to whatever department he or she desires for the teaching of the core courses.

After the UGFE collapsed in 2005, the academic deans, the graduate dean, and the provost held numerous meetings in an attempt to save the UGFE. William "Bill" Boggess, then head of the Department of Agricultural and Resource Economics, and I independently, but at about the same time, concluded that the program should be discontinued. It was our belief the atmosphere surrounding the program was so toxic that its continuation as a productive cooperative enterprise could not succeed. It was more than two years before this was done. At that time, Bill was active in the decision making process at OSU. I was not, having retired earlier.

The death of the UGFE was one of the most disappointing and depressing events of my entire intellectual journey. Its demise deprived OSU economists of an opportunity to provide national leadership in making graduate applied economics a supplement to theoretical graduate economics education at highly elite institutions. And OSU lost an opportunity for the Liberal Arts to participate in a masters and doctoral program. Given the historical constraints on the Liberal Arts at OSU, the UGFE was the first fundamental breach in the system imposed approximately 65 years earlier on the university I love dearly.

LESSONS LEARNED

The UGFE experience was an important piece of the road that I traveled during my intellectual journey. I summarize this part of my journey under three headings. One pertains to the College of Liberal Arts at OSU. Another is the role of applied economics in higher education. The third concerns the UGFE as an experiment in intradisciplinary cooperation.

THE LIBERAL ARTS: From the time I arrived at OSU in 1954 to the present, I have puzzled about the role of the liberal arts in universities that place particular emphasis on an applied science mission. I believe such an applied science mission is best served if the liberal arts are considered full partners. I do not come to this conclusion because I believe this is deserved by the liberal arts a priori. Rather, it is because the liberal arts have a unique and important role in applied science.

What should I have done, then, at OSU, where full partnership did not exist and was not on the horizon? My approach has been to encourage their joint participation in the work of applied science units on campus wherever possible. This is especially true of the social sciences but not limited to them. I do not regret the position I have taken, a position I advanced as dean of faculty and dean of the Graduate School, as well as from other posts I have held.

APPLIED ECONOMICS: I view applied economics as a needed development in the discipline of economics. Graduate programs in applied economics are within reach of many institutions that cannot hope to duplicate well the more theoretically oriented programs at such institutions as Harvard, MIT, Stanford, or the University of Chicago. The participation of economists in the OSU College of Liberal Arts as UGFE partners would have permitted them to do research at whatever level their capacity would permit on subjects in macroeconomics, international trade, industrial organization, environmental economics, and economic development. Nothing has transpired since 1986 to make me believe my convictions have been misplaced.

I was president of the American Agricultural Economic Association in 1972–73. At that time, I became convinced that agricultural economists should become applied economists in both name and deed. I established a broad-based panel at an annual meeting to discuss the subject. The panel discussion made it clear that the membership would not support such a move at that time.

The American Agricultural Economics Association, with leadership from Steve Buccola, has recently changed its name to the Agricultural and Applied Economics Association. It has also created a journal that recognizes the distinctive nature of applied economics research. Three and a half decades is a long time to wait, but it was worth it.

In 1972, the last year I was head of the OSU Department of Agricultural Economics, I requested permission on behalf of the faculty in that department to change the name to the Department of Agricultural and Applied Economics. My request sailed through the College of Agricultural Sciences, but the desk of OSU President Robert MacVicar became an insurmountable boulder. He believed the Department of Agricultural and Resource Economics was more appropriate, and he imposed that name upon us. I was disappointed because it was apparent to me that there was much work the department should do that was neither agricultural nor resource economics. As described earlier, in 1988

I again was instrumental in submitting a proposal to the State Board of Higher Education for graduate degrees in applied economics at OSU. The chancellor's office obviously was unaware of emerging opportunities for applied economics within the economics field and denied OSU the opportunity to be a national leader in applied economics. Despite these earlier failures, I am delighted that in 2009 OSU has seen fit to create a graduate program in applied economics. One is reminded of a statement attributed to Winston Churchill: "Americans usually do the right thing, after they have tried all of the alternatives."

INTRADISCIPLINARY COOPERATION: The UGFE experience can be viewed as an experiment in intradisciplinary cooperation. There is much discussion in academia about the difficulties associated with inter- or multidisciplinary academic efforts. Yet, in this experiment the discipline did not vary. This makes it possible to compare obstacles to cooperation within an academic discipline to those between and among disciplines.

A MULTIDISCIPLINARY EXPERIENCE: THE NATIONAL RURAL STUDIES COMMITTEE (NRSC)

In the summer of 1986, the W. K. Kellogg Foundation requested my participation in a seminar-type discussion about the problems of rural America. I had written an essay on this subject before leaving RFF (Castle 1985) and had let it be known that I intended to direct my attention to problems affecting rural people and places for the remainder of my professional career.

The Kellogg Foundation assembled a diverse group for this seminar. One of the participants was a woman who had spent time in rural places working with nongovernmental organizations concerned with rural problems. She had also worked with the Council of State Governments on assignments that involved her in rural issues. During the course of the seminar, she spoke several times about the contribution that was needed from academic institutions. In essence, she sought a conceptual framework, or classification system, that would permit priorities to be established among the difficulties existing in rural places and with rural people.

As a result of that seminar, I submitted a proposal to the Kellogg Foundation to establish a National Rural Studies Committee (NRSC). The purpose of the NRSC was to provide academic institutions with information that would improve their services to rural people and in rural places. The committee was multidisciplinary. If fresh perspectives were to be obtained, I believed that committee members should come from higher education generally, not be confined to personnel from land grant institutions.

The original NRSC consisted of 12 people other than me, all of whom I appointed. I knew only three of the 12 prior to their appointment. Bruce Weber, my close associate at Oregon State University since 1986, was my first appointee. Edwin Mills, foremost urban economist, was at Northwestern University when he accepted appointment to the committee. I had known Mills for several years but mainly by reputation. Pierre Crosson, economist at Resources for the Future, was known to me as well. I met the other committee members as they were appointed to the committee. These included David Brown, Cornell University demographer and sociologist; Julian Wolpert, Princeton University geographer; Ronald Oakerson, political scientist at Indiana University; Carol Stack, anthropologist at Stanford University; Sonya Salamon, anthropologist at the University of Illinois; Gene Summers, sociologist at the University of Wisconsin; and Edward Bergman, regional scientist at the University of North Carolina. Four regional development centers, a part of the land grant university system, were affiliated with the committee from the outset as well. William Howarth, American literature professor at Princeton University, gave the keynote address at the first meeting of the NRSC. He was much taken with the NRSC and its membership. He attended several additional meetings of the NRSC and became a kind of de facto member of the committee.

The NRSC investigated rural issues and held annual meetings that included field trips to the following locations: Hood River, Oregon; Greenville, Mississippi; Cedar Falls, Iowa; Reading, Pennsylvania; and Las Vegas, New Mexico. Bruce Weber directed a Center Associate Program in conjunction with other committee activities. The target audience of the Center Associate Program was academic personnel who were interested in rural studies and eligible for leave that would permit substantive work on an important rural studies topic. Nine associates studied under the program. In all instances, these people testified to the profound impact of the Center Associate Program on their academic careers. David Brown led a summer session for four-year college faculty held at Cornell University. The NRSC also assembled a 25-chapter, 500-page book of readings on rural America that is extensively cited here. Most of these original essays were prepared specifically for that book (Castle 2005). A complete listing of NRSC activities may be found in the final report to the W. K. Kellogg Foundation, listed among the references at the end of this chapter.

For several years I have taken advantages of numerous opportunities to probe conditions in rural places throughout the United States, especially in the 48 contiguous states. Among these opportunities were numerous automobile trips, often with family, through rural areas, with overnight stops in small rural towns. On these trips, I engaged in conversations with residents about local conditions. These multiple exposures over a long period resulted in general impressions that survived the more intense NRSC experiences.

It is a simple matter to engage rural residents in conversation about local conditions and community welfare. Rural men typically will respond immediately to questions about the local high-school basketball or football team. If women are present, a sure-fire conversation starter is to inquire about local school finance. These people are aware that many of their young have migrated elsewhere for many years. Those with children in school, or with children who will be going to school, generally are concerned with the education their children will receive. Those without children may be concerned with the cost of local public school education, but local people, in one way or another, are interested in local schools. Jobs and employment are another conversation starter, as is health-care availability. It has been suggested that rural people become preoccupied with guns and religion in response to unfavorable rural conditions. My observations provide little support for such conjecture. Hunting and fishing are traditional rural activities, and guns are often used for hunting. Guns are a part of the urban culture as well. Rural crime has increased in many places, but I am not aware it is more gun-related than urban crime is gun-related.

I have also have had opportunities to learn about changes that have occurred in rural communities over a span of years and, in some cases, several decades. I have come to appreciate the importance of what economists label "path dependence" in rural communities. This is the extent to which past decisions in a community affect the community welfare at later dates. I learned that path dependence has been of considerable importance in several communities. Often natural resource conservation and management of health care have been involved. The protection of environmental amenities as economic development occurs has enabled some places to reap economic benefit from those protected natural amenities at a later date. Path dependence with respect to health care often stems from the efforts of health-care workers in an earlier period. For example, a small clinic developed earlier may have been key to attracting physicians or other medical workers to the community later. This, in turn, may permit connections to be made with medical centers in urban places, or at least in larger places. The health facilities in a rural place may be of considerable importance when retirement decisions are made by people in the surrounding countryside.

My Washington, D.C., experience also permitted me to form impressions about rural America. Senators and representatives every four years spend time on the "Farm Bill." Farm bills typically are predominately about the welfare of those who produce and market agricultural commodities. Yet "rural" is more than "agriculture," and from time to time while I was in Washington, hearings or other meeting were held about rural conditions. Often I was invited to attend and sometimes was asked to make a statement or prepare a brief. I came away from such meetings reflecting on the great difference between the state of urban and rural economics. Urban economics was recognized, justifiably, as an established field within economics. There were numerous workers in urban economics, who were supported by much high-quality literature. Rural economics did not have comparable standing because there were inadequate accomplishments to justify such recognition. This circumstance was one of the reasons I invited Ed Mills, noted urban economist, to become a member of the NRSC. He was an exceedingly valuable member. He did not permit us to consider rural issues as disconnected from urban influence. As our work developed, he was able to have rural economics listed as a recognized field parallel to urban economics by the American Economic Association.

The NRSC efforts were carried out over a nine-year period and were financed by three three-year grants. As each of the first two grants approached an end, the Kellogg Foundation invited the NRSC to submit a continuation proposal. Both extensions were financed, but the objectives were changed somewhat for both.

The proposal I initially submitted to the Kellogg Foundation reflected my observation that academic efforts pertaining to rural affairs were diverse in focus and of varying quality. Scholars were scattered across the country, often with little direct communication with each other. Extension workers in the land grant system had programs of varying effectiveness. Many were discontinued when personnel retired. Rural academic workers typically gravitated into the field from other specializations rather than being educated for such work. And, in fact, few graduate programs educated people for work with a rural studies emphasis.

The rural studies field at that time could not be described accurately as a disciplinary matrix (see Appendix C). To cite but two reasons: 1) a commonly accepted, central conceptual framework was lacking; and 2) the graduate education that existed was spread across several disciplines with little in the way of linkage among those disciplines. Whatever commonality existed was not sufficient to provide sustainability.

A new way to think about and describe rural places was needed. Much had been written about how and why cities form and what they contribute to society. We all are better off as a result. Prevailing views of rural places were incomplete. It is well known that rural places provide such commodities as food and lumber and that many urban dwellers have rural roots, although even that is changing rapidly. Rural America is much more than these stereotyped views would suggest. In one way or another, every member of the NRSC expressed a desire to improve this situation.

It was appropriate that an explicit NRSC objective was to provide a conceptual framework for the field. Another was to provide encouragement and assistance for younger academic workers. The purpose of assembling a diverse group of scholars and providing support for younger scholars was to encourage the development of a rural studies disciplinary matrix. The central conceptual framework was of high priority, and it is most appropriate to appraise the NRSC experience from that perspective. The final report of the NRSC is listed in the references for this chapter and permits evaluation of NRSC efforts in this respect. Since then, my thinking has ripened and matured, and two developments have permitted me to go beyond my thinking in 1998, when the NRSC ended its work. One was a symposium held in my honor in 2005 that examined the conceptual frontiers of resource and rural economics in great depth. Some of the results of that symposium are reported in the volume edited by Wu, Barkley, and Weber and published in 2008. The other development has been the writing of this book, which has required me to think more rigorously and creatively about several issues than I had previously. Neither of these two developments would have produced the same results had it not been for the prior work of the NRSC.[3]

RURAL PLACE THEORY

As I was writing this book I decided once again to attempt to formulate a conceptual core for the field of rural studies, which I now prefer to label Rural Place Theory. Five empirical findings follow that underlie and buttresses the theory.

FIVE EMPIRICAL FINDINGS:

- *Rural America is a vast place with a varied landscape populated with diverse human resources.* These diverse characteristics have conditioned local responses to a common, but rapidly changing, external economic, institutional, and natural environment. When particular people or places are considered, their uniqueness is to be appreciated as they accommodate outside stimuli. Yet the accommodation of external influences cannot be totally explained, or socially justified, on the basis of local diversity alone. The rural studies discipline reflects the normative judgment that "it is appropriate rural people have, and exercise, a degree of autonomy in addressing their common concerns and in seeking fulfillment of their aspirations" (Castle 1998, 622).

- *Powerful exogenous economic, social, political, and technological forces affect rural—as well as urban—places and regions.* These forces have power in large part because people generally wish to improve their lot in life; the effects may be different for rural than for urban people more because of different circumstances than different motivations. The contribution of NRSC member Edwin Mills was of fundamental importance in distinguishing between rural and urban adjustments to economic and social change in the United States (Mills 1995). The relation of farming and agricultural to other rural occupations and activities in rural places is of fundamental importance here. There is an unfortunate tendency to use "agriculture" (or "farming") and "rural" as synonymous. It is only early in the economic development process that most rural people are engaged in farming. As development occurs, fewer people farm even though agricultural output increases. With perfect mobility and perfect market adjustments throughout the economy, rural people will migrate to urban employment that is assumed to become available simultaneously. In the United States, young people have migrated in great numbers, especially during the 20th century. Nevertheless, in most, and perhaps all, market-oriented societies—including the United States—poverty and low income are significant rural social problems. Younger people migrate more readily, but as people become older or infirm, they are less inclined to do so.

- *Poverty and low income are found in all rural places but in varying degrees.* Gene Summers, one of the first appointees to and vice chair of the NRSC, deserves credit for directing NRSC attention to, and documenting, the pervasive nature of rural poverty (Summers 1995). He was president of the Rural Sociological Society (RSS) during his NRSC service. After an NRSC meeting in rural Mississippi, he became instrumental in the establishment of the RSS Rural Poverty Task Force. The Task Force developed a Pathway from Poverty project that educated policy audiences in Washington, D.C., as well as regionally throughout the United States. In an NRSC publication, Summers describes fundamental characteristics of rural poverty in the United States. Bruce Weber continued Summers' pioneering work after NRSC efforts ceased. Weber has succeeded in integrating rural poverty scholarship into academic work on poverty nationwide (Weber 2008).

- *Rural land use has profound implications for natural resource and environmental policy.* NRSC member Pierre Crosson identified crucial environmental policy issues stemming from rural land use. He called attention to general environmental impacts of agricultural land use, especially urban impacts. This is a fundamental policy issue in an economy based on private property rights in rural lands (Crosson 1995). "An inevitable tension exists between those who make use of natural resources to produce food, fiber, timber, energy, and minerals, and those who look to the natural environment for ecosystem services, such as natural amenities and recreational experiences" (Castle and Ervin 2008). Segerson believes these tensions come into the open and revolve around the use of, and access to, natural resources in less populated places. Rural America is where most of the natural environment is found, and this is where most such conflicts come into the open.

- *Both centralizing and decentralizing forces are constantly shaping and reshaping the spatial landscape of the United States.* The economic dominance of cities and metropolitan places constitutes a most powerful centralizing force that has not yet run its course, either in the United States or globally. Simultaneously, individuals, firms, governments, and nongovernmental organizations in centralized urban places move outward into less populated areas, seeking advantages in areas with greater or different space (Lewis 1995).

These findings reflect a synthesis of what is known currently about five important conditions in rural America: similarities and differences among rural places, external forces affecting rural people and places, poverty and low income, and rural land use as well as centralizing and decentralizing forces. As useful as these findings are for many purposes, standing independently they do not provide an integrated view of rural people and places. They clearly fall short of providing a better way to view, describe, and serve rural America.

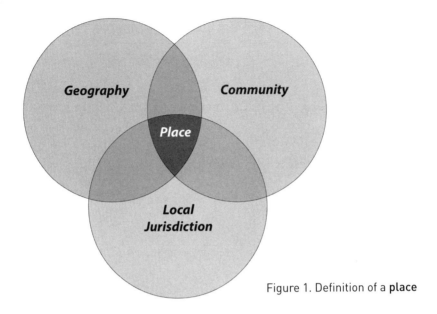

Figure 1. Definition of a **place**

DEFINITION OF A PLACE: Prior to presenting a rural decision model, a definition of *place* is required. As used here, *place* involves the intersection of an identifiable geographic area on the globe, a specific jurisdictional unit of government, and a social community of inhabitants. The Venn diagram in Figure 1 shows *place* as occurring if a jurisdictional unit coincides with a distinctive geographical landscape, together with a human community of interest. Individuals have at least two of these three features in mind when they say something like: "I am from that place" or "what place are you from?" In any case, a geographical component is necessary to distinguish among places with respect to location. There must be some community of interest among people within a place if group decisions are a possible alternative to individual decision making. Finally, access to a formal jurisdictional unit is necessary if local affairs are to be linked to state or federal government.

Sherman County, Oregon, serves as a nearly perfect example of this definition of place. Rivers serve as boundaries for three sides of the county, permitting a specific geographic description. Moro is the seat of Sherman County government and provides access to government generally. The approximately 2,000 people living in Sherman County have educational attainments greater than those of the average Oregon citizen. Their sense of community is demonstrated by the cooperative community project of providing the land, labor, and equipment needed to provide an athletic track and field for the local high school.

A distinct concept of *place* is fundamental in rural studies. As used in rural studies, a *rural place* connotes distance with some geographic space between rural places. To be meaningful, rural places must be sufficiently distinct so that small group decisions and actions associated with the *place* will matter. *Rural places* are a function of space, distance, and relative population sparseness.

A THREE-PREMISE DECISION MODEL: From the above empirical findings, I established three premises that frame a group decision model for people in a rural place.

- *Local comparative advantage, reflecting the natural and human created environment, creates potential for local decisions to reduce gaps between existing circumstances and aspirations.*

- *The U.S. system of government (federal and state) legitimizes a degree of local autonomy in the making of local public policy.*

- *People and places must be considered jointly if rural America is to be understood. Three general orientations of people to a place are possible: 1) inside – outside, 2) outside – inside, 3) inside – inside.* This premise is true by definition but frequently has been overlooked. Most rural people are associated with a place, and how this is reflected in their decisions is of obvious importance. Outside people may affect what happens inside, and their motivations and influence need to be captured. The three general orientations of people to places exhaust all logical possibilities.

These premises permit the accommodation of all decision makers: households, firms, government, and non-governmental organizations (social capital). The three-interlocking-premise model draws heavily on my earlier research concerning local autonomy, comparative advantage, social capital, and intermediate decision making (Castle 1998, 2002, 2005).

THE GENERAL ORIENTATION OF PEOPLE TO A PLACE—SIX EXAMPLE

GROUPS: The three general orientations can now be made more specific by six example groups, with two groups associated with each general orientation. The literature previously cited showed that each example group was of empirical importance.

An Inside–Outside Orientation

- *Group 1.* This group includes those who expect to migrate away from a rural place. This includes the young who intend to have urban employment and lifestyle. This group may include significant numbers who are unemployed or underemployed. Most, if not all, nations at one time or another have had significant numbers in this category.

- *Group 2.* Those who export goods or services from their rural place as their principal source of income are included in this group. Farmers, foresters, and those who commute to urban employment are included here. A higher percentage of the U.S. nonfarm labor force resides in rural places than is true of any other industrialized country in the world.

An Outside–Inside Orientation

- *Group 3.* This group includes those outside a rural place who are attracted to a rural place by natural or man-made amenities or by some combination of the two. These people are attracted to rural places because they provide something not readily available elsewhere. Those institutions that provide access to these amenities are of interest here as well. Rural places that have colleges and universities have flourished economically relative to other rural places for the past several decades.

- *Group 4.* Individuals or groups that seek to undertake economic activities in a rural place constitute this group. Such investment may be of temporary or enduring nature and may be competitive with the amenity desires of Group 3.

An Inside–Inside Orientation

- *Group 5.* This group consists of those who produce goods and services that are largely consumed locally. It includes government employees, shopkeepers, and farmers who sell most of their produce in local farmer's markets.

- *Group 6.* Those who reside in a rural place because of local support systems, either of a market or nonmarket nature, are included in this group. Providers of such support systems include families, religious organizations, charities, local community organizations, businesses, and others. As rural residents becomes older or more infirm, they are less likely to migrate away from rural places.

IMPLICATIONS OF A RURAL PLACE THEORY: At this point, it is useful to review the components of the theory as well as its implications. Five empirical findings serve as building blocks to insure that whatever theory follows is based in realism. The definition of place is grounded in geography, government, and social cohesion. A three-premise model reflects local uniqueness and a role for local government, together with a linkage that joins rural people with rural places. From this it is possible to deduce that there are three, and only three, general orientations of people to rural places: inside–outside, outside–inside and inside–inside. Once these three general orientations are established, examples of specific rural groups, based on previous observations, can be associated with each of the three general orientations (see Figure 2 especially).

The approach just described makes use of central place theory, upon which urban economics and traditional agricultural economics draw heavily. Central place theory provides an explanation for the movement of commodities to urban places, as well as the rural–urban migration of people. In other words, central place theory is concerned with centralizing forces. Spatial adjustments occur constantly in developing economies. Some have centralizing effects—witness the growth of cities and metropolitan places—but some are decentralizing in nature, a subject about which central place theory is relatively silent. Rural place theory provides an integrated view of both centralizing and decentralizing forces. Individuals, firms, and governmental and non-governmental organizations in urban places may look to less densely populated places as a source of opportunities for many reasons. Some may be interested in rural amenities, either natural or man-made; some look to rural places for potential investments; and some may be motivated by lower land rents that translate into lower housing costs. Every place is in some sense unique, if for no other reason than each place has its own location. Yet if that is the only difference among places, market transactions, as assumed by central place theory, will account for such differences with transportation and communication costs. It is in this connection that heterogeneity among rural places comes to the fore. There are various motivations for an "urban invasion of rural America," and individual rural places will not have the same appeal to all who look outward (Lewis 1995); decentralization will be enhanced by the existence of heterogeneous places. By combining rural place theory with central place theory, both flows from rural to urban places and flows from urban to rural places can be accommodated by the same theoretical system.

Those familiar with rural places have long noted the presence of numerous small groups in many rural communities. Figure 2 displays examples of place-related groupings and provides an understanding of why such groups may exist. The basic reason, of course, is that a group may be a more effective way of achieving individual aspirations than individual action decided upon independently. The success of such groups will depend upon whether sufficient trust exists among members to justify expectations of reciprocity. "If I help you rebuild a barn after a fire, will you do the same for me?" Small groups may resolve conflicts within a group and enable them to speak with one voice to outside interests. In many circumstances, groups provide a form of insurance. "Social capital" is an appropriate description of arrangements that provide help when such unexpected events occur.

One should not assume that the actions of all groups held together by trust and an expectation of reciprocal behavior will necessarily be socially benign. At the local level, a group may achieve the equivalent of monopoly control of access to particular resources or opportunities. This may deny opportunities to racial minorities or others who are outside the group. It is for this reason, as well as others, that local groups and rural places are subjected to the rules and values of the larger society, of which they are one part. Rural America has a less-than-exemplary record in its treatment of racial minorities. As noted earlier, I am

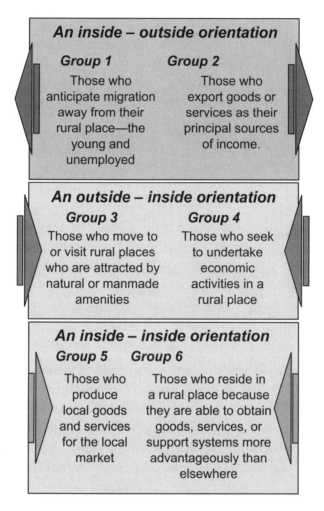

Figure 2. Orientation of rural people to rural places: six groups

fortunate that I did not become imbued with racial prejudices in the places were I was reared.[4]

Decentralization has the potential of destabilizing rural places just as centralization may destabilize cities and metropolitan places. Outside–inside economic processes that require access to amenities, waste disposal, and other resource uses may be in conflict with established inside–outside or inside–inside economic interests. Furthermore, all inside interests may not react in the same way to possible outside–inside pressures. Local merchants, for example, may welcome developments that have the potential of increasing demand for local goods and services. It should not be assumed that outside–inside influences at any given time run their course and then subside forever. For an excellent account of recurring outside–inside influences, see "Conflict and Change on the Landscapes of the Arid American West" by Paul F. Starrs, Chapter 14 in *The Changing American Countryside: Rural People and Places*, E. N. Castle, ed.

The destabilization of rural places and the role of small groups in rural places permit four public policy–related subjects to be identified. These four policy-related subjects are of central importance in rural studies research.

FOUR RURAL PUBLIC POLICY–RELATED SUBJECTS: Use of the three-premise model as a way of thinking about rural issues has permitted me to view traditional rural studies subjects from a broader and more analytical perspective.

- Rural America is a place where standardized, market-priced products that require space for creation (agriculture and forestry) flow to urban places, directed by a decentralized market system. The producers of these standardized products use space associated with a natural environment that yields numerous heterogeneous ecosystem services. The landscape of rural America, then, provides a natural laboratory for the integration of economic and ecological systems; this is where most of the natural environment is to be found. In addition, it is a place where the institutions that serve to integrate the two systems, such as property rights, will be subjected to increasing pressure.

- Variation in the natural environment together with human ingenuity establishes the base for local comparative advantage. Small groups become essential for the creation and exploitation of local comparative advantage; small-group activity is a more effective means of achieving individual ends than individual decisions, individually made (Ostrom). Such intermediate decision making makes use of local social capital, and lies between micro and macro decisions as developed in economic theory. The three-premise model (see Figure 2) serves to identify groups likely to play major roles in local group decisions.

- Underemployment of rural people is common among economies when economic growth occurs as defined and measured by neoclassical economic theory. Some underemployment may be because of economic disequilibrium; some may be a function of the individual because of rural support services and lifestyles. Not even Adam Smith maintained that all people would act as though led by an invisible hand. There is much related to the apparent underemployment of rural people that is not understood. The causes are not the same in every place it is observed; regional differences are great. If it is to be understood, there must be explanations for differences in rewards and penalties across regions. Again, implications of the three-premise model and Figure 2 will be helpful in research design.

- Small-group decisions and local social capital in rural places stand in bold relief to the big corporations and big government organizations that have a presence in even the most remote rural places. Homogeneous, standardized products and services are loved by large organizations and is where their comparative advantage is greatest. Big business and big government exist because they can deliver a standardized good or service at a low cost. Small organizations may have an advantage if the market is relatively small and heterogeneous products or services are required. Rural places, again, may provide a natural laboratory for the consideration of an important social issue. Society may be well served by public and private organizations of varying size existing side by side. "Too big to let fail" may have brought a fundamental social issue into the open that has too long been neglected.

In summary, the approach outlined can be used to establish a rural studies research and education agenda or a rural studies disciplinary matrix. The uniqueness of particular groups within local places and how they are discovered, developed, and treated become an important part of rural studies. External forces such as economic and social change, as well as the aspirations of different groups, become important as well. The reconciliation of local uniqueness with external forces is neither automatic, immediate, or without friction, and may require thoughtful intervention (Castle and Weber 2006). Greater knowledge about the formation and exercise of social capital, and further development of intermediate decision making become an essential part of the agenda.

Currently, rural studies more nearly approaches a disciplinary matrix than when the NRSC came into existence, but more needs to be done. Some, but not all, of the progress that has been made can be attributed to the NRSC. With the help of the above framework, it is not difficult to understand why progress has been difficult. Rural studies workers are separated by discipline, academic boundaries, and geography. Only at Oregon State University, where Bruce Weber has provided leadership, has a campus-wide approach been taken. The Rural Studies Program he directs follows closely the model set forth in the final NRSC report. In the Reflections section of this chapter, I advance a proposal for a way this separation can be addressed.

I now understand much better how, and why, my intellectual journey has led me to this point. I began my career concerned with farm management and land economics. When I first came to Oregon State, my involvement with natural resources led me to the macro tool of benefit-cost analysis. Yet, when I considered resource management from the perspective of the local community and discovered the relevance of local comparative advantage, complications crowded upon me. I learned that neither micro- nor macroeconomic theory accounted properly for the economic significance of local comparative advantage. It was from this that I came to refer to the reconciliation of local and external interests within economics as intermediate decision making.[5]

The following three conclusions follow logically from the rural place theory as developed and applied in the preceding discussion:

• Despite popular discussions that leave the impression that rural America is static, homogeneous, and idyllic, I conclude that rural America is better described as *dynamic, heterogeneous,* and *turbulent.* There is constant change and conflict within rural places as well as between rural and urban interests.

- Rural studies will benefit from the contribution of several disciplines—sociology, political science, regional science, geography, and economics, for example. Yet the organization of most universities makes the emergence of a rural studies disciplinary matrix remote. Nevertheless, the following proposal is doable and merits consideration.

- The insights provided by the rural place theory bring to the fore the possibility of a New Spatial Economics that would bring together central place theory and rural place theory, as well as what is known about the dynamics of edge cities and suburban places.

A PROPOSAL: I believe the field of rural studies will be best served by a coordinated and cooperative approach of several academic disciplines. This would be stimulated greatly by a professional association with a title such as the International Rural Studies Association. The membership should be multidisciplinary. There was or perhaps still is a proposal from some rural sociologists to change the name of the Rural Sociology Association to the Rural Studies Association. I believe this would be a bad idea if this resulted in rural sociologists, or any other discipline, attempting to capture rural studies. A better plan would be for appropriate sections within several existing associations to cooperate in forming an International Rural Studies Association. This could include the Agricultural and Applied Economics Association, the Rural Sociological Society, the Regional Science Association, and other similar groups. If formed, such a group should welcome international scholars because learning about the relation of rural interests to others in the society will proceed best by consideration of more than one nation. A scholarly journal should be established early in the life of the new association, and ways should be found to involve students in meetings of rural studies scholars.

EXTENDED EDUCATION AT OREGON STATE UNIVERSITY

In 1979, a time of considerable student unrest, Martin Trow, a University of California–Berkeley professor, provided a useful description of two higher education traditions in the United States. He referred to one as the *classic, autonomous approach*. This approach provides for the conservation and transmission of high culture, pure scholarship, and the selection, formation, and certification of elite groups. He described the other as the *popular approach*, which provides a place for as many students as possible to continue their education beyond high school and extends useful knowledge and service to groups and institutions.

Clearly, Oregon State University belongs in the popular rather than the autonomous tradition. Even so, it performs functions that Trow associates with autonomous education. Oregon State came into existence as a result of federal legislation signed by Abraham Lincoln that established land grant universities. This system was intended to provide higher education for the sons and daughters of those who lived on the land and those engaged in the "mechanic arts." These institutions pioneered teaching and offering degrees in agriculture, forestry, home economics, and engineering (mechanic arts). Subsequent legislation established the 1890 institutions in line with a "separate but equal" policy, primarily for the African-American people in southern states. (Separate? Yes, indeed! Equal? No way!) In 1914, with the signature of President Woodrow Wilson, the Smith-Lever Act became law. This legislation established a cooperative system of extension education in the United States Department of Agriculture (USDA), the land grant universities, and local governments.

As noted earlier, I returned to OSU in 1986 and served as chair of the University Graduate Faculty of Economics (UGFE) for six years. By that time, OSU was suffering from a series of budget reductions, the UGFE was functioning as intended, and I was nearing 70 years of age. I decided to retire as UGFE chair, although I continued to work part-time. During those years, OSU President John Byrne was making every effort to organize and operate the university in an increasingly austere financial environment. He obtained the assistance of an outside consulting organization to suggest changes that might bring this about.

In 1993, Cooperative Extension was administered by the College of Agricultural Sciences. Extension had personnel in every county in Oregon, as well as considerable staff in colleges of Agricultural Sciences, Home Economics, and Forestry. For several years, some Cooperative Extension administrators and faculty had maintained that they should be located administratively in central administration rather than in the College of Agricultural Sciences. Interestingly, this was not opposed by College of Agricultural Sciences administrators. Oregon State University had the good fortune at that time to have had the most progressive cluster of higher education agricultural administrators in the nation. That cluster included Roy Arnold, Thayne Dutson, and Conrad "Bud" Weiser.

The report of the outside consulting group to President John Byrne was discussed and evaluated by numerous groups within OSU prior to implementation. As this process was nearing an end, President Byrne invited me to his office and said something like the following: "Emery, we have rearranged things at OSU. Everything fits fairly well except that we have one part left over, and we do not have a place for it. That part is Cooperative Extension, and I hope you will help me find the right place."

John Byrne and I had known one another for many years. We were OSU department heads at the same time and served on the committee that helped make OSU a Sea Grant university. I considered his request a compliment, but I had no preconceived ideas about an answer. He said he would compensate me for my efforts. I replied that I would need to reflect on the problem to be addressed, but there would be no cost involved even if I concluded I could be helpful.

And reflect I did. Had this been a simple issue, John Byrne would not have needed assistance. Many changes had occurred in society and in Oregon's economy since Extension first became operative. The organization was led for many years by Director Frank Ballard, who built one of the most highly regarded Extension programs in the nation. With the passage of time, the agriculture and forestry industries were revolutionized. Related issues transformed rural households and rural family life. County Extension personnel at one time were repositories of, or had ready access to, information of great practical value. But their role then began to change. The needs of people on the land and in rural households became more complex. Many came to know university scientists who had the latest scientific information at their fingertips. Such developments did not make local Extension personnel obsolete, but they did affect how Extension could be most effective. There was ample confirmation that some Extension personnel knew they needed the knowledge base provided by the university. As evidence, one needed only to consider their wish to become a university-wide program rather than be confined to agriculture, forestry, and home economics. Arguably, rank-and-file Extension personnel understood the university better than rank-and-file faculty within the university understood Extension.

The surface issue concerning the OSU Extension Service, if it became a university-wide organization, was to what central office it should report. This was the problem on John Byrne's mind when he invited me to his office. Should it report directly to the provost or should it report to a vice president who had responsibility for campus-wide centers and institutes? As I reflected on this matter I realized this decision had ramifications for Extension personnel on campus as well as for those in the countryside. When I returned to John Byrne's office to discuss his request for assistance, I told him I visualized this as a top-to-bottom issue and not just a matter of recommending a reporting assignment in central administration. He accepted my analysis but said he could not wait forever for my recommendations. I do not recall whether he specified how much time I had to report back to him. I do recall I soon began work and gave high priority to the preparation of a report.

The "givens" in the assignment were that Extension would be in central administration in the university at one extreme, and in numerous communities, especially counties, at the other. Given the extremes, there was need to provide specific guidance between these two extremes about its location and governance, both on and off campus. There were two obvious sources of information I could tap in formulating a recommendation on these matters. One was the experience of other states. The other was the judgment of others, both in Oregon and elsewhere. I conferred with people in the USDA who were in a position to know about the administration and performance of Extension across the land. From such sources, it was relatively easy to assemble references and other secondary data. I also then conferred widely with approximately 90 people, both individually and in groups (Castle 1993c, 26). The question was whether Extension should be parallel to academic units such as colleges and departments or integrated into those units. A review of arrangements across the land revealed there was precedent for both. The experience elsewhere was reviewed on a case-by-case basis, because conditions affecting Land Grand universities in the various states are variable, and the land grant universities themselves are very different one from another.

And then a great break occurred. The dean of the College of Agriculture volunteered the assistance of one of his staff. Gwil Evans, one of the best writers on campus with experience in Extension, reviewed my work at every step and then edited what became my final report. Gwil's great judgment, good humor, and technical expertise repeatedly improved the report and my morale.[6]

After collecting information from secondary sources as well as information from individuals inside and outside the university, I concluded, and then recommended, that Extension personnel should be integrated into academic units within the University. There were two fundamental reasons for this. One was efficiency. A parallel organizational structure would have duplicated many administrative personnel at various levels within the university. For examples, think of budget and publication offices. The other reason was better education, both internally and externally. The integration of Extension with resident instruction and research held the potential of improving the subject competence of Extension work. It also provided potential for greater realism and relevance in resident education and research.

There were objections to an integrated path from both Extension personnel and resident faculty. Some Extension people, especially younger people, were supportive. Others feared they would be at a competitive disadvantage when rank and tenure decisions were made. Extension personnel at OSU had long had academic rank, but with integration there would be direct competition and comparisons within academic units among research, teaching, and Extension personnel. On this issue, the great leadership of Conrad "Bud" Weiser came to the fore. He became an advocate for the belief that scholarship could be demonstrated in many ways in addition to the printed page. This belief was officially accepted and integrated into salary, promotion, and tenure decisions at Oregon State for research, teaching, and outreach activities. Other criticism came from resident Oregon State faculty who believed their academic prestige would suffer if their academic units included Extension personnel. Fortunately, this academic elitism was not taken seriously by most faculty or most administrators.

My written report to John Byrne recommended that extended education become integrated with resident education and research at Oregon State University. The federal funding supporting Extension activities was such that most outreach work continued to be within agriculture, forestry, and in the newly created College of Health and Human Sciences (which combined the colleges of Home Economics and Health and Human Performance). Nevertheless, the attention of every college on campus was directed to off- as well as on-campus educational activities. By making the Extension Service a campus-wide unit, in combination with other outreach activities, and making extended education a responsibility of every academic unit, in principle the organization of the total university was consistent with the land grant university ideal.

When John Byrne received my report, he released it publicly with a two-page statement of his supporting decisions at the outset of the report. This was a courageous act because he was risking criticism from both university personnel as well as from across Oregon. He then assigned implementation of his decisions to university Provost Roy Arnold. I have said that Roy Arnold's was the most demanding of the three assignments and that John Byrne's decisions were a close second, with the preparation of my report being the easiest of the three.

My report was given to John Byrne in June of 1993. The journey of extended education since then has not been smooth sailing. It would have been unrealistic to have expected otherwise. Some extended education personnel have had to become more familiar with traditional academic values and traditions. Perhaps a greater adjustment has been required of traditional academic types with little knowledge of the philosophy and practice of extended education. Surveys have been made of the reaction of extended education personnel after a decade or more of experience with the integrated system. Not surprisingly, satisfaction with the existing system is inversely related to length of service. Younger faculty who have more recent academic education are relatively better satisfied with the reorganized system. In summary, there is little evidence that any significant number of affected people wish to return to the system that was in place prior to my report to President John Byrne.

One of the fundamental problems associated with placing Extension personnel in academic units pertains to promotion and tenure decisions. This problem existed prior to my report, but the implementation of my report brought the problem into bold relief. Extension personnel are engaged directly in addressing problems that arise in society. They do not have the luxury of confining their attention to the elaboration of an academic discipline. This is true of many applied scientists as well. Yet it should also be recognized that the establishment of a common base for comparison among faculty is not confined to faculty in residence and those in Extension. Similar issues arise if the same criteria are used to evaluate faculty in music, art, and literature with, say, those in chemistry, physics, and biology.

REFLECTIONS[7]

The reader will be justified in asking if the three activities reported on in this chapter have anything in common except that they were performed by me after I returned to OSU in 1986. I maintain they do, indeed, have much in common, but their commonality is not immediately obvious. All were concerned with the tension that develops in academic institutions from attempts to serve both autonomous and popular functions in higher education. My experience, reported here, occurred at Oregon State University, a land grant university. Yet the central tension reflected in each of the assignments is fundamental. It is likely to be present in any higher education institution that attempts to address autonomous *and* popular functions.

In the first example, concerned with the University Graduate Faculty of Economics, the issue was whether a single-discipline graduate education program could serve both traditional theoretical graduate education as well as applied educational objectives. In this case, the answer would be no, unless priorities were established in advance, or unless a decision process could be agreed to in advance and then respected. This example also served to demonstrate that sharp and deep conflicts can and do occur within academic disciplines. This should be

remembered when the potential problems of multidiscipline cooperation are being considered.

The second example, the experience with the National Rural
Studies Committee, demonstrated the advantage of a multiple-discipline view of complex social phenomena. If the autonomous functions
of the university are to be served, the integrity of recognized, traditional disciplines must be respected. Yet disciplinary integrity need not
be sacrificed when certain popular educational needs are addressed.
The National Rural Studies Committee membership reflected several
academic disciplines—geography, regional science, economics, political science, anthropology, sociology—and academic appointments in
prestigious universities. In this case, a complex social issue was analyzed far more satisfactorily with multiple-discipline cooperation than
would have been possible by independent activity of representatives of
the same disciplines.

The third example—the question of where extended education
activities at Oregon State University would be placed within the
organization of the university—involved dealing with concerns that
autonomous academic pursuits would fail to serve popular needs. From
another perspective, there were concerns that problems arising in the
Oregon countryside (popular needs) would distort resident research
and education priorities at OSU (autonomous academic pursuits).
These are important and valid concerns. In the minds of many, these
concerns have been addressed in the organization that has emerged.
Nevertheless, we should not lose sight of the ever-existing tension
between these valid activities.

The three examples can be compared from the "disciplinary matrix"
perspective as advanced by Kuhn (see Appendix C). In science, according to Kuhn, a necessary condition for the formation of a matrix is a
discovery sufficiently interesting to attract others to develop implications of the discovery. Graduate programs, textbooks, academic journals, and professional associations are stimulated by such continuing
efforts. Heroes, icons, and traditions arise. The University Graduate
Faculty of Economics was based squarely on mainstream intellectual
discoveries in neoclassical economics. Clearly, some other ingredient
was missing rather than a common intellectual discovery. The obstacles

to cooperation stemmed from differences between "normal science," as described by Kuhn, and applied science. As explained elsewhere, normal science stems from a disciplinary discovery; applied science derives from problems in society.

The multidisciplinary National Rural Studies Committee members enjoyed remarkable cooperation for nine years in developing and elaborating a field of rural studies. There was no single intellectual discovery to unite workers in this field even though the field itself became better defined during the committee's existence. In this case, an important social problem was sufficient to hold the group together for nine years. There are continuing rural studies efforts that stemmed from the original NRSC initiatives.

I turn next to the reorganization of extended education at OSU. In this example, difficulty arose because, in one case, there was a tradition of service to people regardless of location. This conflicted with a different tradition: that traditional quality education occurs in a place where a community of scholars is located and where scholars have autonomy in their service to this tradition.

The lesson that emerges from the three experiences is that lasting academic endeavors are held together by different types of glue. One is the intellectual glue described by Kuhn that results from a revolutionary discovery. Another glue stems from the nature of social problems that call for cooperative intellectual efforts. Still another glue is derived from the repeated use of a particular methodology. A particular culture often develops around the methodology, and the heroes in a field are those who have been especially successful in using the methodology. The lesson to be learned here is that the administration of academic endeavors requires that attention must be given not only to subject-matter competence, but also to social practices and cultural traditions. Such considerations were discussed in Chapter 4, concerned with graduate education.

7

THE LAND GRANT IDEA, THE IDEAL, AND THE 21ST CENTURY
(1942–2010)

In the fall of 1942, I entered Kansas State, my first exposure to a land grant university. Since then, applied science and land grant universities (LGUs) have been central to my intellectual journey. Even when I was at the Federal Reserve Bank of Kansas City and Resources for the Future (RFF), I was involved periodically with land grant issues and people. And, of course, the RFF agenda involved the application of science to public policy, an activity of the LGUs as well. This chapter permits a retrospective appraisal of my LGU experiences and includes thoughts about the future of LGUs.

My undergraduate education has served me well in many respects, even though it was not especially rigorous. Basic science courses were taught side by side with applied science agriculture courses. Science was applied on farms both on and off campus. This was done easily because both Kansas State and Iowa State had farm enterprises on campus and access to farms off campus. I knew little about LGUs as institutions, but I soon learned that scientific knowledge and its application were to be taken seriously. And, as noted previously, thanks to Milton Eisenhower at Kansas State I came to appreciate the liberal arts. I never questioned the relevance of the cultural history and philosophy courses that were part of my agriculture curriculum. I also became aware of extended education programs in agriculture that were associated with LGUs, although extension people had never visited my family's farm. I knew the Extension Service served farmers and believed it might be a source of employment when I obtained an agriculture degree. In summary, my undergraduate education had breadth and was never far removed from practical affairs.

I did not fully appreciate the uniqueness of the LGU approach to higher education until I was at the Federal Reserve Bank of Kansas City. When I was on the outside looking in, I became conscious that the land grant approach rested on the application of dependable scientific knowledge. I wanted to be a good economist and concluded then that I wanted to apply what I knew and what I might learn to practical affairs.

LAND GRANT UNIVERSITIES, APPLIED SCIENCE, AND EXTENDED EDUCATION

Land grant universities are a uniquely American institution. They were created during the Civil War when the nation was less than a century old and reflected certain values of the young country remarkably well. Founding fathers John Adams and Thomas Jefferson, as well as others, placed the birth of the nation on a firm intellectual base. It was fitting that support from the federal government would be forthcoming for education that was especially suited for those doing the heavy lifting on farms and in small towns. The total package provided unique dimensions: Scientific knowledge was to be applied to important problems, and education was to be extended in places where people were to be found—on farms and in factories and in the kitchens.

Despite these and other commonalities among the various states, LGUs are far from homogeneous. In some states, the land grant program is to be found at the principal publicly supported institution in the state. The University of Wisconsin provides an example. Cornell University, in contrast, is a private institution, and a member of the Ivy League, with four state-supported statutory or contract colleges. In other states, the LGU came into existence mainly to provide for the administration of the LGU legislation. Oregon State University provides one such example; Kansas State and Iowa State provide others. In summary, the LGUs vary in size and in the roles they play within their respective states, as well as in academic prestige.

The LGU designation often creates a stereotyped image of the educational programs that are offered. For many years, the term "cow college" was applied to some LGUs, although currently less frequently and more selectively. As they have evolved, the LGUs began to offer degrees in numerous fields and disciplines. The LGUs vary widely in how they interpret and embrace the land grant mission. In some cases, this mission is reflected in university-wide goals and objectives. In others, the mission has become mainly the responsibility of colleges of agriculture, forestry, and home economics, with perhaps some contribution from engineering. At Oregon State, I believe it accurate to say that the land grant mission is taken seriously in much of the university.

It is useful to compare the mission of land grant universities as they were envisioned when created with the way they have evolved: ". . . without excluding other scientific and classical studies and including military tactic, to teach such branches of learning as are related to agriculture and the mechanic arts. In such manner as the legislatures of the States may respectively prescribe, in order to promote the liberal and practical education of the industrial classes in the several pursuits and professions in life" (The Morrill Act, 1862). Although a distinctive and particular mission was provided for the land grants, the liberal arts and sciences were not neglected or excluded. Subsequent legislation provided special support for research and extended education consistent with the above mission.

In some instances, the land grant mission has been meaningfully applied to important contemporary problems. In others, there is a nearly mechanical exercise of traditional LGU activities and mere lip service to such ideals. Mechanical maintenance of traditional programs results in an increasingly narrow focus; serious application of the LGU approach requires constant innovation and change.

In view of this extreme heterogeneity, it is difficult to make valid generalizations about the entire LGU system. For many purposes, it is more fruitful to think of land grant research and education as an approach, or as an idea, rather than in particular institutional terms.

The land grants are lauded and copied in various places around the world. Their commitment to applied science and education, both extended and resident, deserves careful evaluation.

One need not confine oneself to agriculture, forestry, and home economics when evaluating the land grant approach to education. Sea grant universities have been a reality since the mid-1960s. The first sun grant universities have been named. The development of health grant universities could address many important problems. The following model is useful if one wishes to think about merits and demerits of the land grant approach.

COMPONENTS OF AN IDEALIZED LAND GRANT UNIVERSITY MODEL

- *Resident Education.* This provides resident higher education for those who wish to do professional work in a particular arena. For example, the health industry requires a wide range of human talent in addition to physicians and nurses. There is increased emphasis in much resident education in providing a research experience for undergraduates. This is consistent with the traditional land grant philosophy.

- *Extended Education.* One of the most original features of the traditional LGU operation was to extend education to people in places where they lived or worked. In many cases, knowledge becomes more powerful if taken to where it can be applied (see Chapter 6).

- *Applied Research.* The breadth of existing and potential LGU research is significant. Social need is a difficult taskmaster, and the anomalies that arise in applied science can, and do, lead to the most basic of research. Decentralized research facilities permit differences among places to be considered in research program design and application.

The unique feature of the three components is their interlocking nature. Each of the components can be found in various places in educational systems, but the distinctive feature of land grant education is the linkage that exists among the three components. It is important that all three components be kept in mind when the land grant model is discussed.

THREE TASKS: I have had three tasks that required evaluation of the LGU approach to research and education. A brief description of each follows.

- *The Commission on University Goals.* James Jensen served as president of Oregon State University from 1961 to 1969. During this period, he set in motion several developments that had not come to fruition in 1969, when he announced his intention to resign. Following this announcement, considerable striving for position developed among various units within the University. It became clear that an acting president would serve for several months, perhaps for a year or more. Jensen took action to stabilize conditions on campus with the creation and appointment of the Commission on University Goals. The three people constituting the Commission were relieved of other responsibilities sufficient to permit them to devote half of their time for one and a half years to this assignment:

> To assist Oregon State University in the development of a clearer definition and understanding of its purposes and goals.... In order to carry out its primary responsibilities of assisting the institution develop recommendations for long range planning, the Commission will be expected to study and evaluate existing organizational structures and methods of operation, present and emerging programs, to determine how effectively they serve the purposes of the University. The Commission will be responsible directly to the President, and it will be free to seek information and opinions from any office or individual in the University.

C. Warren Hovland, chair of the Department of Religious Studies, James Knudsen, assistant dean of the College of Engineering, and I were named members of the Commission (sometimes referred to as the "Three Wise Men"). This assignment required that we consider how the resources of the University could be used to best serve the aspirations and needs of OSU students and Oregon citizens. This assignment caused me to appreciate the complexity of a modern university and better understand the purpose for, and relation between, fields of knowledge and units within a land grant university.

- *The National Association of State Universities and Land Grant Colleges: A Commissioned Lecture.* When I was at Resources for the Future, the W. K. Kellogg Foundation provided funds for a commissioned lecture to the Council of Presidents of the National Association of State Universities and Land Grant Colleges (now the Association of Public and Land Grant Universities) at its 1980 annual meeting. An Association committee chose the person to present the lecture based on submitted outlines of what would be covered in the lecture. I submitted the successful outline. The Ford Foundation then made a special small grant to me to cover expenses associated with assembling a few small groups to provide the information I wished to have. The lecture was given at a time when there was controversy about how land grant university research was funded and administered. My lecture examined the rationale underlying different points of view as to how this should be done (Castle 1981).

- *Oregon Boards and Commissions.* Over a 25-year period, I served on such bodies by appointment of five Oregon governors. Historically, Oregon has not had a typical cabinet form of government. Boards and commissions have had considerable power and influence as a result of serving in an intermediate position between the state legislature and the governor's office. I have been a member of the State Water Resources Board, the Water Policy Review Board, and the Environmental Quality Commission. Such bodies deal with policy issues arising from the administration of state and federal legislation. This service is described here because frequently it was necessary to draw upon technical information and personnel available in colleges and universities. This permitted me to evaluate public services provided by universities from an "outside looking in" perspective.

LAND GRANT UNIVERSITIES AND THEIR STAKEHOLDERS

THE TRADITIONAL VIEW: In the Kellogg lecture referred to earlier, I made use of a 1979 *Science* article by Evenson, Waggoner, and Ruttan entitled "Economic Benefits from Research: An Example from Agriculture."

This article permitted me to begin to untangle the complex relationship between land grant agricultural research, the interests of society, and the agricultural industry. The following summary statement is taken from the *Science* article:

> . . . that the increase in farm labor productivity is outstripping the increase in nonfarm productivity: that farm productivity is rising significantly even though the economy has stagnated; that public investment in agricultural research has yielded relatively high rates of return, ranging from 20 to 90 percent in 32 studies reviewed; and that agricultural research and extension are significant contributors to the productivity of American agriculture.

The authors described the agricultural research establishment as having three distinguishing characteristics—articulation, decentralization, and undervaluation. They provided evidence that the nation is poorer because of its failure to invest more heavily in the system. The best scholarship at that time indicated that the agricultural research establishment is, indeed, highly productive, based on existing market prices and government policies. Articulation within the system permits close communication links between those who develop and those who use technology. Decentralization permits place and location to play explicit roles. Research is planned and conducted in experiment stations, branch stations, and substations. Agricultural research in the land grant system illustrates that the productivity of science is motivated by social problems. It also demonstrates the unique contribution of applied science to basic research.

The importance of an abundant and inexpensive food supply should not, and need not, be discounted. The percentage of wealth and income spent on food in the United States is the lowest any population has ever enjoyed. Life spans have increased even as this has been accomplished.

The above quotation from the *Science* article brings into the open the characteristics of the traditional land grant university. In particular, a highly productive research and educational system, if measured within the context of market prices, and a relatively homogeneous clientele.

THE CURRENT SITUATION AND PROBABLE FUTURE—UNINTENDED CONSEQUENCES AND A HETEROGENEOUS CLIENTELE: There are potential dangers in a decentralized research system designed to serve a particular industry. Will applied scientists in such a system inevitably become captured by the clientele they serve? If so, will the values of the clientele trump other social objectives? The returns to agricultural research identified in the cited article were based on market prices and existing government programs. Environmental side effects, either positive or negative, were not taken into account. I have had a ringside seat from which to observe how the land grant research establishment has incorporated environmental considerations into its agenda in the years since the above-cited *Science* article was written.

Although not so described when I first went to OSU in 1954, economic development generally was more "politically correct" than environmental protection. We were (and are) fortunate at OSU to have a Department of Fisheries and Wildlife in the College of Agricultural Sciences. I had long been aware that economic theory stipulated that either positive or negative environmental externalities could arise in market-based economic systems.[1] My service on Oregon boards and commissions provided opportunities for me to learn of and deal with both. These opportunities and experiences provided me with a more balanced view than was available to many others in the land grant system.

In 1966, Nyle Brady, director of the Cornell University Agricultural Experiment Station, under the auspices of Section O of the American Association for the Advancement of Science, assembled a symposium in Washington, D.C., on the subject "Agriculture and the Quality of Our Environment." This is also the title of the proceedings of the symposium, which were made available a year later in book form by the Academy as AAAS publication 85 (Brady 1967). The symposium focused on the impact of agriculture on the environment and the effect of the environment on agriculture—specifically, air, water quality, soil pollution, and human and animal waste as related to agriculture and the environment. So far as I am aware, this was one of the first systematic attempts to address environment–agriculture relationships. Most of the participants were biological and physical scientists; I was the only social scientist to give a formal paper, and as I recall, there were only two or three social scientists at the symposium. Even so, I was impressed by the substantial number and the quality of the participants.

Since the first Earth Day in 1970, attitudes have changed. Environmental protection has become more politically correct than economic development. Yet the LGUs cannot be proud of their failure to record, identify, measure, value, and research the environmental effects resulting from the use of the production-increasing technology they helped create. I regret I did not do more to persuade agriculturalists to give greater attention to the subject.

Prior to Earth Day in 1970, it was my good fortune to assist in helping resource economics become a recognized field in economics (Castle, Kelso, Gardner 1963; Castle 1967). I believe this recognition occurred because RFF, OSU, the University of California, and other institutions involved recognized economists, some of whom became Nobel Laureates, in pondering economic anomalies arising in natural resource economics. Their help was sought because accepted economic theory and quantitative procedures did not work well when used in actual resource management situations. In my experience, those scientists at the revolutionary edge of their fields are more likely to be interested in difficult applied problems than are those preoccupied with routine or normal science activities.

The colleges of agricultural sciences within the LGUs typically are unapologetic about their need to serve their stakeholders. This is understandable because stakeholders both provide funds directly and persuade others to provide support. Credibility may never become an issue unless other members in society are affected negatively or the research is challenged in the public policy arena for other reasons. When this occurs, the publicly supported research program can no longer be preoccupied with a single, relatively homogeneous, group of stakeholders; the research it conducts has entered the public policy arena.

Land grant university administrators now know the traditional model is no more. A homogeneous clientele no longer exists; multiple viewpoints must be taken into account. Both intended and unintended consequences need to be considered. Even though the operating environment has become more complicated, the maintenance of credibility is no less important than it was previously, nor will it be less important in the future.

As I reflect on the problems I have faced as an academic administrator and in special assignments, I am struck by how many of them involved unprecedented situations. (Some of these are described in earlier chapters.) Reflection causes me to conclude that most methodological literature deals with the expected rather than the unusual. I have little original to add to the extensive literature on research in a stable institutional environment. The usual assumption in the research methodology literature is that the institutional framework is stable. The environment in which research is conducted often is far from stable. The viewpoint adopted in this section is that of an academic administrator who wishes to assist those in society who seek solutions to "messy" problems and to enhance, or at least maintain, the credibility of his or her organization in the public eye.[2]

"MESSY" PROBLEMS AND CREDIBILITY: Messy problems typically are persistent because they are *controversial* and *complex*. Parties of interest often seek different outcomes. One reason they seek different outcomes is that they do not agree on what constitutes *means* and *ends*. Complexity may arise for a number of reasons, including a poorly defined institutional environment, disagreements on what constitutes reality, or great uncertainty combined with high stakes for one or all parties. The following paragraphs provide a discussion of terms and issues I have found to be important when dealing with "messy problems."

- *Collaboration:* Some researchers and research administrators may wish to avoid collaborative relationships with parties of interest. This is understandable, especially if there is more than one party of interest and if a formal collaborative relationship is under consideration with fewer than the total number of parties. Under such circumstances, questions of bias may arise resulting from a conflict of interest. *Nevertheless, collaboration as such is not necessarily cause for alarm.* Perhaps the ultimate in a collaborative relationship is the joint involvement of land grant university researchers and farmers in decentralized experiment stations. There may be merit in involving parties of interest in certain matters because it may be in their interest to identify relevant, but otherwise overlooked, information. If a controversial messy problem involves multiple parties of interest, it will be exceedingly important to be transparent in dealings with one and all.

• *Multiple Viewpoints:* One of the attractions of universities in messy problem investigations is that a variety of disciplines and expertise is present in the university. Controversy often stems from different perceptions of reality. If different perceptions of reality can be isolated, it may be possible to identify needed research within various disciplines. With many messy problems, progress may result from narrowing the areas of controversy rather than the discovery of a grand solution. During more than two decades of service on Oregon boards and commissions, I do not recall anyone calling for an optimal solution or the maximization of an objective function as an answer to a messy problem. Progress may be made simply by reaching agreement on the elimination of socially inferior alternatives. The final result may then emerge from negotiation and political decision making.[3]

• *Crucial Information:* Parties of interest with knowledge of the messy problem may provide important information even though they may be biased. Early on, I learned to pay attention at public hearings when I served on Oregon boards and commissions. Such testimony is often uninformative and repetitive. Yet I cannot recall a single important controversial issue for which valuable but previously overlooked information was not supplied by public hearings. When a messy problem is first approached, one often does not know what one does not know. One way to overcome such ignorance is to *listen* to people with knowledge of the messy problem. I personally interviewed 90 people when I prepared the extended education report for John Byrne described in Chapter 6. I am confident I avoided many missteps by doing so.

• *Assigned Responsibility with Commensurate Autonomy, Authority, and Accountability:* The research administrator involved in messy problem resolution should seek assignment of responsibilities with commensurate autonomy and authority. By doing so, problems can be avoided that may arise in the conduct of research or services. For example, it may be desirable to have an agreement at the outset as to how changes in the administration of a project may be made after it is under way. I once had a collaborative funder who wanted to redefine the research problem when he became dissatisfied with preliminary research findings. Even though the results of applied messy problem investigations may not be published in normal science outlets, normal science peer-review processes often can be used to advantage. Furthermore, messy problem results may make important contributions to the normal science literature.

- *Credibility:* Applied science research organizations vary greatly in the credibility they enjoy. Credibility is exceedingly valuable but typically is earned in highly subjective ways. It is part and parcel of the reputation of any organization. High credibility usually is acquired over a considerable period of time, but it can disappear rapidly. And, once lost, it is not easily regained. When I went to Resources for the Future, it enjoyed high credibility earned over the previous 25 years. This reputation was based on: 1) published research that was highly regarded by scientific peers, 2) nonpartisanship, and 3) nonadvocacy. When I became president of RFF, I concluded that its credibility was its most valuable asset and that credibility must be maintained even though the cost of doing so might be high. At one time, we were close to financial disaster, but even when financially lucrative opportunities arose, they were not pursued if we believed they posed a risk to our credibility.

An obvious threat to credibility pertains to the source of funds supporting research. Are LGU research results to be trusted if the research is financed by the industry most directly affected by research results? What about other funding sources? Do public foundations ever attempt to influence the results of research they fund? Do government agencies always keep their hands off the research process once a research project is funded? The correct answer to these questions leads to the conclusion that funders have credibility reputations as well. The skillful applied science administrator will have guidelines regarding the relation of funders to the research process and results obtained.

It is helpful to think of messy problems in three categories. Firstly, those problems with a high probability of being resolved in a relatively short time. A second category comprises those with an uncertain date of problem resolution but which may be on a path to resolution. The third involves those with no discernible prospect of problem resolution; they may be a problem to be managed rather than solved.

As land grant Universities have matured, the public policy implications of doing applied science research have become highly significant. Not only have problems to be researched arisen in society, the results typically will be of interest to more than one group. Like it or not, research results must withstand public scrutiny.

MESSY PROBLEMS—THREE EXAMPLES

Descriptions of three messy problems follow. I have been involved in research and educational efforts related to the first two described. My interest in the third, which is concerned with climate change, stems from my interest in scientific methodology and as a concerned citizen.

AGRICULTURAL POLICY: My first exposure to agricultural policy was as a member of a Dust Bowl Drought and Depression family during the 1930s. New Deal farm policies were pragmatic efforts to provide financial relief to farm families at that time. There were resource management requirements that had to be respected in order to qualify for payments. Those programs were forerunners of what exists today, but two basic forces are reflected in today's farm policies. One is the way democracy works in practice. The other is the transformation of the agricultural industry itself. Agricultural production and marketing have been revolutionized in my lifetime. Farms now purchase many more inputs, usually in the form of new technologies, that previously were provided, if at all, on the farm. Commercial fertilizers, improved seeds, and premixed livestock feeds are a few examples. And many more marketing functions are now performed off the farm than was previously the case, such as preservation of foods, packaging, and penetration of distant markets.

Adam Smith had it right. Declines in the cost of production are closely related to the extent of the market. Farms have grown in size but not as rapidly as firms in many other industries. Technologies have contributed greatly to increased farm output, due in no small part to LGU research and education. Many of these technologies took advantage of, and some were the result of, low energy prices. Capital, in the form of technology, was substituted for labor and land. Young people left rural for urban places in great numbers; land moved from the control of small and medium-sized farms to those of much larger size.

Agricultural price supports for some agricultural commodities that began in the 1930s have been continued and increased. In industry after industry, bigness became the norm, and the agricultural industry has not escaped criticism. The safety of food supplies has been questioned, and negative environmental effects of agricultural production have been noted. There is speculation about whether political coalitions can continue to be formed that are strong enough to maintain agricultural subsidies. Low incomes and poverty exist in all rural areas (Summers 1995), but agricultural price supports are not an effective way to address these conditions. For several decades, most rural people have not farmed, and most rural income has not resulted from farming (see Census of Agriculture).

Even the largest of farm firms are not nearly as large as the largest industrial and financial service firms. In my opinion, *bigness* is a neglected dimension in public policy discussions. Antitrust policies do not address bigness as long as competitiveness is maintained. Yet increasingly we hear that some firm is regarded as "too big to let fail." This means that my welfare—and yours—is diminished if the firm in question grows beyond a certain size. When the failure of "too big to let fail" firms looms, government action is usually of two types: government "bails out" the firm by the injection of government funds, or the firm is merged with another "too big to let fail" firm. Neither action addresses the "bigness" issue as such.

Competitive market theory assumes large numbers of relatively small firms. Under such circumstances, the failure of one firm, or even several, will not impose significant economy-wide costs. When competitive conditions do not exist, measures that will discourage firms from becoming "too big to let fail" should be given consideration. In financial services, reserve requirements could be made increasingly stringent as firms increase in size, especially as they approach being "too big to let fail." For other industries, bankruptcy laws could apply regardless of size, and, if bankruptcy occurs, assets could be sold only to firms that are not "too big to let fail." Such preliminary thoughts deserve elaboration, debate, and empirical testing.

The largest farm firms typically are not as large as the largest non-farm businesses that sell inputs to farmers and purchase, process, and market farm products. Nevertheless, farm output is becoming increasingly concentrated. Historically, many agricultural programs were justified on the grounds that large numbers of farmers benefitted from such programs. With increases in size of firms generally, it is becoming increasingly difficult to justify many agricultural policies on the grounds that large numbers of farmers are benefitting. Justification arguments for many programs, including justification for subsidized agricultural research programs, has shifted to low-cost food for consumers. Everyone must eat, and benefits are widely dispersed. Traditional agricultural policy is no longer a headline public policy agenda item, either publicly or within agricultural economics.

ENERGY AND THE ENVIRONMENT: As noted in Chapter 5, the decade I spent in or near the nation's capital permitted me to observe the Carter and Reagan administrations from close range. Not surprisingly, I was especially interested in their energy and environmental policies. At least since the 1930s, petroleum supplies have been curtailed from time to time. During the Ford presidency, just prior to my going to Washington, D. C., global oil supplies were significantly reduced. From that time to the present, every President has given lip service to energy independence. This pious hope does not serve the nation well, for two reasons. First, our economy and lifestyles would be drastically affected if this were to be achieved except over an exceedingly long period of time. I have seen no evidence that we are willing to make this adjustment, nor has any President advocated necessary measures for doing so. Second, turning a searchlight on energy without directing attention to energy-environment interconnections neglects subjects that need attention, including the economic realities of any energy-environment related policies. Economic activity and lifestyles are heavily dependent on inexpensive fossil fuels. There are other sources of energy available, but they are expensive or environmentally questionable. Understandably, the environmental movement has not directly and aggressively attacked economic growth as a threat to the environment.

The Carter administration expended considerable political capital on energy and was generally friendly to the environmental movement. The Department of Energy was created during the Carter years. Near the end of Carter's term, the energy problem became less serious for several reasons. The cumulative effect of both consumer and producer adjustments to higher energy prices was of significance. Further, petroleum exporters came to recognize it was not always in their interest to squeeze the best customer in the world for petroleum products.

Chapter 5 has a discussion of contacts I had with members of the Reagan administration. Initial Reagan appointees considered in some circles to be unfriendly to the environment were replaced during Reagan's first term with appointees more acceptable to the environmental movement. Even so, there was considerable tension between the large environmental organizations and the Reagan administration for its entire tenure.

It was my observation that the first Bush presidency improved relations with environmentalists significantly. This was due in no small part to the appointment of William K. Reilly as administrator of the Environmental Protection Agency. Reilly was president of the Conservation Foundation prior to its merger with the World Wildlife Fund. He and I came to know one another when he was president of the Conservation Foundation. From all accounts, he and his President (George H. W. Bush) had a most harmonious relationship.

There are many more battles to be fought in this messy energy-environment war, but there are promising developments. Energy use per unit of GDP is declining. Energy conservation is absolutely a way to go. Yet much, much remains to be done. Neither environmental nor economic growth objectives will be well served unless their deep interconnection is well understood. The great need is for public policies that cause these interconnections to be taken into account in the choices citizens make in every facet of the economy. This objective will not be served by energy uses that diminish human health or destroy the natural environment.

Energy use has both direct and indirect effects on the environment. The direct effects are environmental consequences that can be attributed to an energy source, such as the amount of carbon residue that is created by burning a ton of coal in a particular way. There may be important indirect effects as well. Assume that an energy source is discovered which has no direct environmental consequences and which can be made available at no cost or at a very low cost. It is reasonable to expect that this would be a great boon to economic growth. If so, that growth would likely have enormous environmental consequences, not all of which would be benign.

Fundamental here is the attitude of those concerned about economic growth and the environment as to how and to whom the implementation of environmental policy is to be entrusted. The historic answer to this question has been government. For seven years, I sat on Oregon's Commission on Environmental Quality and made decisions that implemented federal and state environmental law. The legislation typically defined desired implementation in terms of practices to be followed. For example, if it was desired that fish were to inhabit a given stream, practices were specified that would be met quarter by quarter each calendar year. In other words, environmental requirements were stated both in terms of the desired result and the methods or practices needed to bring such results about. This is typically described as the "command and control" approach. This approach often has brought about good results, so long as technology is changing slowly or not at all.

Different approaches have been proposed. Many economists advance proposals that would make it in the interest of private-sector decision makers to make more environmentally benign decisions. Consider the consequences of fossil fuel use and its effect on the atmosphere. Global warming aside, the carbon that is spewed into the atmosphere is neither healthy for humans nor attractive to behold. A carbon tax on the use of fossil fuels might result in reduced use but, perhaps more importantly, provide more private-sector incentive for the development of technology that would provide substitutes for fossil fuel use, or, when used, would make the effluent more benign. With either approach, both government and the private sector have roles to play.

This discussion pertains to one of the greatest of human paradoxes.

It is a search for principles to guide relations among humans, as well as with the natural environment. Fundamental to this search is the recognition that we cannot know the consequences of our actions far in advance. Programs, policies, plans, and actions surely will be subject to change as new information emerges. The capacity for adaptation will become of obvious importance. Surely this attempt will be enhanced by the employment of many minds in contrast to vesting responsibility with the very few.

CLIMATE CHANGE, GLOBAL WARMING, AND A PHILOSOPHY OF SCIENCE: The possibility of climate change caused by carbon dioxide emissions stemming from fossil-fuel use has changed the public debate over energy and the environment. Possible drastic and irreversible climate change has tended to become the reason for doing something about the appetite for fossil-fuel use. Yet taking carbon from underground and releasing it into the atmosphere is not a healthy thing to do from a number of perspectives. Conversely, those who have great concern about reducing the enormous use of fossil fuels, which they believe is essential to economic growth, direct attention to what they call the "flimsy science" of human-induced global warming.

I am not an atmospheric scientist. I believe I understand the arguments advanced by scientists on both sides of this question, but I am not qualified to dispute either. Nevertheless, Thomas Kuhn has taught me about scientific revolutions, and that encourages me to offer observations about the controversy. In the first place, nothing is ever settled in science. We may judge that scientific knowledge is sufficiently reliable to support particular public policies or individual decisions, but if it is really science, we do not know anything with absolute certainty. I prefer to characterize climate change science as neither "settled and certain" nor "flimsy." From this perspective, it is appropriate both to argue for public policies that will curtail the discharge of carbon into the atmosphere and to question the scientific basis for such policies. But such arguments will neither allege that they are based on scientific certainty nor assert that there is no scientific basis for a counterargument.

Kuhn tells us that two conditions must exist for a scientific revolution to occur. First, anomalies, or conditions not explained by existing theory, must exist. Second, an alternative theory must be advanced that provides at least a theoretical explanation for the anomalies that have been observed. And, even when these conditions are met, the older theory does not necessarily disappear.

As I think about climate and the possibility of man-made warming, I do not believe an alternative theory to the role of carbon in climate change has been advanced to explain the anomalies that have been observed. And I do believe there are anomalies. Even so, the presence of anomalies does not necessarily mean that human-induced carbon in the atmosphere has no effect on the climate. But that is far different than arguing that the matter has been "settled" from a scientific point of view. Rather, that argument directs attention to such matters as probable outcomes and the costs of wrong decisions.

I do not know the motivations of all parties to this controversy. It is not difficult to imagine scientists and others on both sides who are motivated by large financial gain or significant recognition. Kuhn tells us that it is normal for some scientists to be protective of prevailing theories or discoveries. Transparency of method, sharing of data, and peer reviews are devices to insure that anomalies and new concepts are considered. Yet followers of the most recent discovery often dominate the scientific "mainstream." It is my observation that courage, persistence, and good luck may all be required to crack the mainstream control of peer-reviewed publications in some situations. And, in the situations I have observed, the stakes have been puny compared to climate change issues. In any event, skeptics perform an essential service if science is to be served.

I favored a carbon tax on fossil-fuel use long before the climate-change controversy arose. I believe our economy and our lifestyles are much too dependent on an energy supply that has been available over time at a declining real cost, a use that assaults humans and the rest of the natural environment, directly and indirectly. I regret the nature of the climate-change controversy, because it is being waged on grounds that may damage perceptions of how scientific questions get resolved in science. Regardless of how it turns out on scientific merit, we may be distracted from thoughtful consideration of a more fundamental issue—the interrelation of energy, economic growth, and the environment, with climate constituting only one part of the environment.

REFLECTIONS

THE LAND GRANT IDEAL: Earlier in this chapter, I itemized the components of an idealized land grant university model. Beyond the tightly integrated programs of extended education, applied science, and resident education within the land grant universities, it is difficult to make the case that the land grant ideal is widely practiced. Yet there is considerable evidence that the land grant *idea* is alive and well. In other words, the idea that problems arising in society are worthy of attention by educational institutions is widely accepted. Yet, other than in land grant institutions, tightly integrated programs of applied research, extended education, and graduate and undergraduate resident education have not been widely adopted.

The vast majority of the colleges and universities in the United States have outreach missions they take seriously. Four-year colleges service specific educational needs of community residents throughout their lives. Even the most prestigious universities in the nation typically have programs that serve the needs of particular groups and address contemporary problems. Furthermore the LGUs themselves have become flourishing and successful multipurpose universities. Excellence in the autonomous educational pursuits does not disappear necessarily with the offering of popular educational activities. My experience has led me to believe that the land grant idea has not outlived its usefulness. That same experience tells me that at least part of the land grant idea has become integrated in U.S. higher education.

No doubt there are good reasons why the integration of resident education, extended education, and applied science research—the land grant university ideal—has not been adopted more broadly. Nevertheless, the existing land grants will, from time to time, make decisions about their futures. Often such decisions stimulate strategic planning exercises that direct attention to possible reorganizations. Too often, possible reorganizations are studied and debated in the absence of a common understanding about objectives to be achieved or problems to be solved. The remainder of this chapter focuses on the identification of opportunities that will be available to some land grants if they wish to reevaluate their role in society.

The legislation creating the land grant system of education was based on a reading of the need for education by a young, growing, and predominately rural nation at that time (1862). With the benefit of hindsight, most scholars have concluded that the need was not only assessed correctly, but that the means that were created have worked remarkably well. That is to say, a significant social need was identified, and the consequences of the action taken were, on balance, desirable. It is significant that the land grant legislation did not bypass, ignore, or downgrade traditional classic and scientific fields of study.

At this juncture in this book, and at this stage in my career, if I am to be intellectually honest, I must ask and then address some hard questions: Is there justification for continuing to pursue the land grant idea, or the land grant ideal, in public policies? If, in fact, the land grant idea has been adopted widely, is there reason for arguing that the land grants should continue to receive special attention in public policies and with special appropriations? If the land grant ideal—an integrated program of resident education, extended education, and research—has not been copied widely, should attempts be made to establish such an ideal more broadly?

Based on my experience, I believe there is public benefit to be realized from applied science university research on controversial and complex subjects. I also believe there is benefit to be realized from combining such research with related educational activities. Yet not all applied science research justifies resident or extended education, and I favor combining such activities with research as the need arises. I base this conclusion on my land grant and RFF experiences. Such research is useful in public policy decisions, and universities are sources of multiple-discipline expertise.

The land grant university ideal of combining programs of research with resident and extended education requires special justification. The point of view taken here is that when land grant or other comparable universities decide to make plans for the future, they should strive to establish a purpose comparable to the scope of the original legislation. What social issues likely to dominate the 21st century are comparable in scope to those that faced a young rural nation in the mid-19th century? The global interrelations that now exist are of obvious relevance to such issues and questions. In the paragraphs that follow, I set forth several personal observations on this matter.

Three important social goals occur to me as comparable in breadth to the original land grant objectives for an applied science educational undertaking. One goal is *human health*, with particular attention directed to the quantity and quality of food. Another is the *ecological health* of natural systems in the countryside, where most of the natural environment is to be found. A third is *economic health*, especially the economic health of the countryside, including but not limited to agriculture, forestry, and fisheries.

- *Human Health.* As related to food, human health takes on new meaning when considered in an international context rather than if one's thinking is limited to the United States. Even so, there are significant areas of overlap. If the international scene is considered, it is probable that human populations will continue to increase well into the 21st century. This means there will be many hungry and unhealthy people unless enormous innovation and adaptation occur. This is familiar territory to the land grant universities, although population growth is an arena that has not been emphasized in recent years. Abundant food does not guarantee healthy people, of course, but it makes a necessary if not a sufficient contribution.

- *Environmental Health.* It is in the rural areas that most of the natural environment is to be found. Furthermore, colleges of agriculture, forestry, fisheries, and related scientific fields typically are strong in the biological sciences. Protection of the natural environment and resource use for commercial purposes promises to be of great importance in the future. Not surprisingly, environmental conflicts typically arise in rural areas. The rural areas, both domestically and abroad, provide a natural laboratory for the study of integrated ecological and economic systems.

- *Economic Health.* The LGUs have long been concerned with the economic health of particular groups. There is now need to refocus these efforts on the rural countryside. Traditionally, the educational and research efforts of LGUs have enhanced the flow of goods and services from the rural countryside to the cities and metropolitan areas. More recently, attention has been paid to activities that go in the other direction—that is to say, activities that are stimulated by urban interest in the countryside (see chapter 6). The amenities related to space and variations in nature, as well as goods and services stemming from human ingenuity, are of great importance for economic health, as rural places generate their own dynamics in the presence of economic growth. Institutions necessary for accommodating greater rural-urban interdependence cry out for attention. I envision a program that would also reevaluate the role of agriculture in economic progress. Few, if any, economies have been able to avoid stagnant rural economies and rural low income as agriculture has become more efficient. This process needs to be better understood (see Chapter 6 for the development of Rural Place Theory). I can envision a "Healthy Countryside Program" that would necessarily involve those concerned with a healthy environment and those concerned with a healthy economy.

The three objectives previously set forth—human health, countryside economic health, and environmental health—are interlocking. This means their interrelationships must be considered rather than viewing them as independent and in isolation one from another. Begin with countryside economic health and environmental health. In the first place, poverty is no friend of the environment. Unless certain economic essentials are met, little effort likely will be devoted to preserving the environment. Conversely, unless the environment is protected, it will be difficult to maintain economic activity. By the same token, economic health and human health are mutually reinforcing. Thus, at a general level, reflection makes it clear that there are interlocking objectives. Nevertheless, beneath the surface, human created, or social, institutions need attention that will direct the three objectives to be reinforcing rather than antagonistic. Consider, for example, a farmer producing grain for export to a central place, a typical inside–outside producer who sells to a central market, as described in Chapter 6. Next, think about urban-based individuals or groups that are interested in the rural land that produces grain for export to a central city, land that also has the potential to provide rural amenities to those in urban places. The question then becomes whether a system of property rights can be developed that will satisfy both interests. Rural place theory, developed in Chapter 6, makes clear that the simultaneous integration of centralizing and decentralizing forces determines the kind and level of countryside economic activity that will exist. The details of this integrated economic activity will influence the short- and long-run environmental health of the countryside, as well as the welfare of people who have interests there.

The previous paragraph is of potential significance to those who administer land grant universities. Traditionally, the land grants had a special interest for those who live on the land, mainly farmers and farm families in a wide range of circumstances. As agriculture has changed, this interest has shifted gradually to larger commercial farmers. There is no need to be apologetic about, nor to sever, this historic relationship in order to recognize and embrace emerging needs. It would be exciting, indeed, if some land grant universities were to commit to an economically and environmentally healthy countryside populated by healthy people. This would constitute a contemporary objective comparable in scope and ambition to the original land grant purpose.

It will be immediately obvious to those familiar with land grant universities that existing organizational structures in the land grants did not originate to address the above goals. It would be surprising if it were otherwise. Two observations are relevant. First, reorganization for a purpose is very different from reorganizations that proceed without common understanding of the purpose to be addressed or the problem to be solved. Second, one of the reasons I felt at home at Oregon State when I first arrived in 1954 was the ease of working with the various disciplines and across academic units. That characteristic also played a role in my decision to return to OSU in 1986, when I relocated from Washington, D.C. To let existing organizational structures interfere with work on important social problems is, again, to put the cart before the horse.

The call here is for research programs that are grounded in applied science and for resident undergraduate and graduate education programs that would provide knowledge of human, environmental, and economic health, and their interrelationships. And special attention must be given to the educational needs of those separated from major metropolitan places for reasons of space and distance. I call for the land grant approach to be refocused and applied to the most burning issues likely to arise as the 21st century unfolds.

CORNERSTONES FOR EXCELLENCE: I have long maintained that one of the principal problems afflicting many land grant universities is what I label the Ivy League Envy Syndrome. This syndrome requires that excellence can be measured only in comparison with institutions created for a different purpose and with different resources at their disposal. The syndrome is made worse by popular rankings of the educational merits of different institutions. Little harm may come from the presence of the syndrome in large, well-financed universities that also happen to be land grant institutions, such as the University of California at Berkeley, the University of Wisconsin, and Cornell. Yet there are others, such as Oregon State, Kansas State, and others, where the syndrome may exist, where the cart has been placed before the horse. Permit me to explain.

The first step on the road to excellence in any educational institution is to decide on the social need it can serve for which it has the highest comparative advantage. The three goals set forth earlier are examples of realistic goals that will permit any land grant institution to do credible research with associated resident instruction and extended education. Yet excellence in performance is not an easy taskmaster. The unique requirements of excellence associated with these goals demand careful thought as well.

Applied Science: Land grant universities should aspire to excellence in applied science. Basically, this means that scientific effort will be directed to the solution of problems that arise in society. It does not mean there necessarily will be no theoretical or basic scientific research or scientists on campus. It certainly does mean that applied scientists on campus will know about developments in basic and theoretical fields related to the applied research they are doing. I have long maintained that there is a strong connection between fundamental or basic science and applied science—a stronger connection, indeed, than exists between applied science and normal or routine science. As I have noted elsewhere in this book, it is my experience that many of the most accomplished basic and theoretical workers are "turned on" by difficult applied science problems.

The determination of monetary incentives and other rewards for those engaged in applied science is of great importance. If rewards for outstanding applied science are determined on the same basis as rewards for basic or theoretical studies, it should not be surprising if applied workers begin to be afflicted by Ivy League Envy Syndrome and lose sight of their mission. There is nothing mysterious going on here. If my job description is different than my reward system, I am likely to conclude that my job description is window dressing.

Credibility. The credibility of research units within universities is of growing importance. There are numerous research organizations outside of universities, staffed by competent people, that will undertake a wide range of investigations with results packaged as a private rather than as a public good. If universities are perceived as being "captured" by particular groups, their "public goods" may have little or no advantage relative to the private goods that can be obtained elsewhere. I know from experience that some university research lacks credibility for that reason. If a land grant university were to pursue the three goals identified in the previous section, numerous possible clientele groups would be involved. Multiple clientele groups probably are more conducive to the establishment and maintenance of credibility than a single, relatively homogeneous, group.

Credibility is earned over time, and the ways it is earned were discussed earlier. Transparency in purpose, method, data, and arriving at conclusions are the essentials. Political correctness and the vested interests of collaborators cannot be perceived as relevant if credibility is to be maintained.

The Philosophy of Science and Land Grant Universities. My readers to this point will know that I have spent all of my career in organizations concerned with applied research. The research output from these organizations was intended to be useful in private and public policy decision making. I have not been involved directly in significant research aimed at obtaining "knowledge for knowledge's sake." Nevertheless, some of the research I have been involved in has made minor discoveries that have advanced knowledge quite apart from its immediate application or use. Further, I have had administrative responsibility for researchers who were doing research of a basic, rather than an applied, nature.

Perhaps the greatest surprise and disappointment to come from these experiences is the general lack of knowledge about the philosophy of science—by both basic and applied workers—in the organizations in which I have worked. This lack of knowledge has existed both with regard to general knowledge of the subject as well the applicability of this literature to the conduct and administration of applied science teaching and research. If rank and file land grant workers were to learn and apply the growth of knowledge literature, the effectiveness of the system would be improved enormously, because many costly mistakes would be avoided. There are a few people with whom I have worked who have improved the work of those around them by orders of magnitude. I refer to such people as F. E. Price, James Jensen, Roy Arnold, Theodore Schultz, and Gilbert White. They were exceedingly sure-footed as they traversed dangerous academic terrain, and in all cases the steps they took were consistent with principles to be found in the philosophy of science.

Should one wish to search for it, evidence may be found to support either a pessimistic or an optimistic view of the likely future contributions of land grant universities. The Association of Public and Land Grant Universities (APLU) may provide a window to future LGU contributions. The forward-looking activities of the Board on Agricultural Assembly are of particular interest. The Food Systems Leadership Institute, partially financed by the W. K. Kellogg Foundation, recognizes the interdisciplinary nature of food, nutrition, and health. Opportunities are provided to younger workers in land grant universities to acquire advanced training in leadership skills, institutional change, and food systems generally. The work of the Institute is consistent with a land grant initiative set forth earlier in this chapter that addresses human health, ecological health, and the economic health of the countryside.

8

REFLECTIONS OF A PRAGMATIC ECONOMIST

DISCOVERY AND DEFINITIONS

When my former student Dan Bromley learned I was thinking of describing my intellectual journey in book form, he sent the following, written by Thorstein Veblen in 1898:

> The economic life history of the individual is a cumulative process of adaptation of means to ends that cumulatively change as the process goes on, both the agent and the environment being at any point the outcome of past processes. His methods of life today are enforced upon him by his habits of life carried over from yesterday and the circumstances left as the mechanical residue of the life of yesterday.

I reflected many times on this quotation as I struggled through the previous seven chapters. Giving deep meaning to Veblen's words describes my intellectual journey accurately. Writing the previous seven chapters was a discovery process for me. I began writing with the belief that the philosophical position I now hold had been forged over a lifetime of intellectual effort. I have changed that belief as a result of the struggle I have had in writing the previous chapters. I discovered that my present philosophy was largely in place by the time

I had completed my undergraduate education. Later study, reflection, and experience caused earlier beliefs to be refined, perfected, made more academically respectable, but not changed significantly.

But why should anyone care? I had never intended to discover and advocate the *one best* way to conduct research or administer academic institutions. I do not believe any such thing exists. Even so, I thought "going in" that my experiences would interest those struggling with similar issues and be of some value, perhaps, to those who might take similar footsteps. I now believe the lesson to be learned is different. If we understand, "going in," that experiences facilitate learning, then interesting possibilities crowd into our minds, rather than the crippling assumption that we are just repositories of external happenings. Learning may indeed facilitate the emergence of a philosophical orientation; in my case it enriched orientations I already held. I have come to the view that experience and learning are intertwined and inseparable. The Veblen quotation should be read with that possibility in mind.

My choices and decisions have, for a long time, been consistent with a pragmatic and pluralistic philosophy. The following short descriptions of both will help readers know what I am writing about and why I have chosen to do so.

- *Pragmatism.* The essence of pragmatism is that practical results count. It rests on the notion: we think so that we may act. And the practical consequences of our actions provide tests of the beliefs that motivated us to take actions. At the opposite end of the spectrum is a theory or conception that has no regard for practical consequences. While pragmatism is not based in theory, it does not rule out the use of theory if theory helps in discovering the practical consequences of actions.

- *Pluralism.* I set forth here two versions of pluralism. One version holds that reality consists of many entities rather than an organic whole. With this belief, different viewpoints are required for investigations of reality. If many entities exist, it is unlikely a single viewpoint will illuminate all entities equally well. Another version maintains that reality consists of an organic whole so poorly understood that a single viewpoint is not a fruitful way to pursue an understanding of reality. A superior way is to employ multiple viewpoints in investigating parts of reality.

As a practical matter, I believe pragmatism and pluralism belong together. A pragmatist, but not a pluralist, would be doomed to trial and error after attempts to test a hypothesis based on the single scientific theory available. A pluralist, but not a pragmatist, would not be required to consider the practical consequences of any approach to various theories. Clearly, this would not serve the applied scientist well, since he or she necessarily is concerned with problems arising in society.

TRANSFORMING EXPERIENCES
The following brief summaries explain how my experiences have caused me to become a pragmatic pluralist who makes use of scientific concepts and theories.

THE DROUGHT AND DEPRESSION YEARS (1929–1939): These years cover the 6th through the 16th years of my life. My two older brothers reached adulthood during that period. They went into a world of 15 to 25 percent unemployment with only a high-school education and a few months of business college. My family extended charity to others even when our own health and survival were in question. My parents were innovative and imaginative as they coped with a hostile natural and social environment. Mom hatched and marketed chicks locally and by mail order before that became an accepted commercial practice. Dad rented a farm that had been foreclosed by an insurance company, then planted clover on the farm and pioneered a swine enterprise on the open fields. Not all of their innovative efforts were successful, of course.

Over the decade from 1929 to 1939, average annual Kansas rainfall was 10 percent less than the long-run average. In three of these years, rainfall was 10 percent less, and in two of them, there was more than 25 percent less rainfall than the long-run average. The economic depression caused the price of farm commodities to plummet. About the best economists could do was to observe that prices might be increased if production were reduced even more. Government programs were implemented that took land from cultivation, and livestock were slaughtered.

I had little basis for thinking about, or analyzing, the exogenous forces that were causing the hardships my family experienced. I discovered later that the mature elite were not much better off in their understanding of what had gone wrong or what to do about it. There was little evidence that policy makers drew on established bodies of knowledge as they fashioned programs to help desperate people. Even so, I formed the impression that my government did want to help and was trying to find ways to do so.

My family experienced long working days, drudgery, lack of medical attention, and little intellectual or cultural stimulation during those years. Clearly, I was permanently affected by the experiences of my childhood and teen years; conditions that prevailed from my 6th through my 16th year influenced my behavior throughout my life. This included the way I met responsibilities as a professor, applied economist, and academic administrator.

WORLD WAR II: The emotional impact of the Drought and Depression was soon followed by personal experience in World War II. I observed firsthand the capacity of the Army Air Force to change tactics rapidly in response to unanticipated events. I noted this first when I was in training, when adjustments often were made to make the best use of whatever item was in the shortest supply. This was driven home forcefully when I was in combat as well. There, tactics changed rapidly in response to battlefield experiences rather than holding rigidly to predetermined positions.

The December 31, 1944, mission to Hamburg was described in Chapter 2 and need not be repeated here. As noted, the losses were exceedingly heavy in our bomb group and significant in the Eighth Air Force generally. We were without fighter protection on that mission, and the experience there provided information about the effectiveness of the formation we were then using. As noted earlier, a bomb group consisted of 36 planes. We flew in three squadrons labeled lead, high, and low. On that day, we were leading our bomb group as the lead plane of the lead squadron. We did not fly for a few days after that mission because we had to replace planes and crews.

Soon after we returned to flying missions, a different formation was ordered. The new formation consisted of four, instead of three, squadrons. With the new formation, a squadron had nine, rather than twelve, B17s. Apparently the heavy losses that occurred when we were without fighter protection provided an empirical test of the effectiveness of the formation we had been using. The firepower of a B-17 was considerable and could bring down fighter planes if hits could be scored. The problem was that the fighter planes were far faster and more maneuverable. The new formation increased the exposure of attacking fighter planes to the gun turrets and other 50-caliber machine guns on the B-17s. All of the B-17 crew members I spoke with believed the change was an improvement.

I knew little about the planning and technical considerations that went into the decision to change formations. Yet coming as it did, it appeared to be a pragmatic decision taken in response to the loss of bombers to fighter attacks. I was impressed that a large and, in many respects, rigid and inflexible organization could make an important change in tactics rapidly and efficiently based on solid empirical information. No doubt a great deal of information was taken into account in the decision making that I did not know about. Nevertheless, a common-sense, pragmatic decision was made promptly, with enormous consequences. The Drought and Depression affected my attitude toward life generally; a World War II decision may have literally saved my life.

KANSAS STATE UNIVERSITY: My decision to attend Kansas State after World War II was opportunistic rather than pragmatic. Nevertheless, two educational experiences occurred there that have affected my intellectual journey in fundamental ways. Both were described in Chapter 3 and need not be elaborated greatly here. One was the course in applied economics taught by Professor Ed Bagley. In retrospect, I now believe I was starved intellectually for conceptual, theoretical, liberalizing knowledge about my environment that had affected me, my family, and friends so greatly. Bagley demonstrated the practical relevance of economics with the tools provided in economic theory. Theory was made practical, and I realized the theoretician could have great leverage in affairs of the world.

As mentioned in Chapter 3, the course Man and the Cultural World, which I took as an undergraduate, also had a great impact on me. The course dealt with the literature, music, art, and politics that existed in particular historical periods. Recitation classes were with small groups of students. My recitation class was taught by a historian, Professor Wilcoxsin. He was adept at integrating insights from different fields of knowledge. I took the course just as the Cold War was beginning and there was great concern about a nuclear war. One day, a classmate asked Wilcoxsin if a nuclear conflict with the Soviet Union was likely. His reply was in the form of a mini-lecture that integrated knowledge from a number of fields and academic disciplines. His conclusion was that conflict was unlikely, but that we were likely to live during a long, dangerous period of great uncertainty. I left that class thrilled with his performance and speculating whether I would ever have the knowledge and ability to discuss an important problem as he had.

In retrospect, I now recognize that the basic philosophy that was to guide me for the remainder of my life was largely in place by the time I obtained my baccalaureate from Kansas State. The elements of pragmatism were acquired as a result of the external environment that existed during my childhood and youth, together with the response of my parents and government to those conditions. My World War II experience further supported my pragmatic tendencies. My undergraduate education taught both the importance and intellectual beauty of the conceptual world. And Professor Wilcoxsin demonstrated that multiple points of view were necessary for serious attempts to understand reality. My pragmatic orientation was made more powerful by the conceptual and the theoretical and further enriched by pluralistic points of view.

My friend and colleague Steven Buccola has not challenged my self-described philosophy of pragmatism and pluralism. He has, however, required that I face up to the fact that such a philosophy is morally deficient. That is to say, it requires that nothing be stipulated about what ends are to be sought. He notes, "pragmatism's supporters' claim that contemplation alone is adequate for grasping any truth cuts us off from the depth of the world outside of ourselves." He goes on to say, "I agree with you pragmatism is the best approach to applied science. For other aspects of our experience, we must pursue non-pragmatist ways of knowing."

Steve has directed my attention to two features of my pragmatist and pluralist philosophy that are of great importance. I feel comfortable with one, but the other requires more thought and reflection than can be given to it here. In the first instance, pragmatism and pluralism place a great responsibility on the person who takes those terms seriously. They imply considerable freedom in the choice of theory and method as one seeks to understand reality, because the ultimate criterion is what works in practice. Old theories are not necessarily replaced by new ones; all theories are in some sense partial and incomplete. The companion to freedom is responsibility—a responsibility I have found exhilarating. I like the idea I am not obligated to assume that the price of an asset always reflects all available information, or that government always provides the best remedy for a negative externality.

I am not comfortable with the emptiness of pragmatism regarding relationships among people and the role of truth and goodness in those relationships. Because of Steve Buccola, I now recognize why, from the outset, the title of this book has been *My Intellectual Journey*. I have never considered pragmatism to be the moral basis of my relationship to other individuals. I now understand better than I did why that choice was made and why I may want to address the issue in the future, although not in the pages of this, or any, book.

FORMAL STUDY AND CREDIBLE SCIENTIFIC EVIDENCE

FORMAL STUDY: Although I now recognize that my philosophical base was in place when I finished with my undergraduate work, one should not conclude that I was comfortable with what I knew about the philosophy of science. The economic theory I read when I began my masters degree and conversations with colleagues made me aware that, within economics, a body of literature existed about which I knew little. Earl Heady's research methodology course encouraged me to dig deeper in the philosophy of science. Even so, my most productive formal study did not begin until 1967, when I began to teach research methodology.[1] This continued until 1975, when I went to Resources for the Future. When I returned to OSU in 1986, I again taught research methodology. The literature by then was much more extensive, and I was able to explore numerous issues that had troubled me for many years.

Several economists and philosophers have given attention to methodological problems in economics. Caldwell, Blaug, McCloskey, and Hausman provide examples. I learned that a new way of viewing an issue does not require that one reject everything one had learned previously. One need not consider oneself a logical positivist, or a strict follower of Karl Popper, to make use of those contributions. Such literature had been, and continues to be, useful even as I worked to make pragmatism and pluralism operational.

Economics had great appeal for me because it provided answers, at least in theory, for many troublesome social issues that are addressed only indirectly in the philosophy of science. Deterministic economics permits the economist to deduce conclusions from models with assumptions to the effect that logical people are motivated by self-interest. Yet when one asks if the world really works that way, things become more complicated. And special problems arise when attempting to incorporate pragmatism into deterministic models. Pragmatism questions the notion that a sharp distinction can be made between means and ends. A pragmatist may react to the phrase "education for life" by saying "education *is* life." I reflected on the means-ends dichotomy when discussing strategic bombing near the end of Chapter 2. Some consider strategic bombing to be solely a *means* of winning wars, or bringing about a desired outcome or a favorable result. Others view strategic bombing of civilian populations as an undesirable result in itself, independent of any other result it may have. In one case, strategic bombing brings about a desired outcome, an *end* achieved. In the other, strategic bombing, when used in a particular way, is itself an undesirable *end*.

Economists and other scientists, themselves, often provide examples of how groups within a society may view means and ends differently. At several points in this book, examples have been noted of why scientists are attracted to doing normal science. They often become preoccupied with building and maintaining the instruments needed for investigations within an academic discipline or as followers of a major scientific discovery. They are attracted to making empirical measurements that relate to their specialty as well. "Puzzle solving," made interesting by their theory, may fascinate them also. Such activities are, in a very real sense, *ends*, or desirable outcomes, to them. Nevertheless, in a larger sense such activities are *means* to an *end*, rather than an end in itself, to those in society who wish to use scientific findings for practical purposes. I have concluded that one can be both a pragmatist and an economist, yet one cannot be a naïve pragmatist and a naïve economist simultaneously. I define a naïve pragmatist as one who operates without theory and a naïve economist as one who operates without regard to empirical consequences.

Such issues have been a source of stimulation for me as I have addressed real-world issues in my research and have administered applied research programs. Economics provides a powerful analytical tool for addressing real-world problems. Yet there is great challenge associated with understanding the applicability of scientific findings. This may be the reason I have never been satisfied focusing exclusively either on theory or application for an extended period. I believe formal study and deduction have added greatly to the power of the philosophical position I derived from early life experiences.

CREDIBLE SCIENTIFIC EVIDENCE: References are often made in the popular press to speculations that given public policies will, or should, be based on science rather than "politics." If there is elaboration, the term "credible scientific evidence" may be used. Presumably this means an applied scientist will be able to provide credible scientific evidence relevant to a burning public policy issue. Whether this will occur and be useful likely will depend on whether the affected parties have the same understanding of credible scientific evidence.

Credible scientific evidence is defined as a *consensus* among recognized scientists to the effect that a particular finding or hypothesis describes reality better than any alternative that has been investigated. How can it be known that a consensus exists? There is more than one possible measure. Expert testimony is one. Agreement among the textbooks in a field is another. Peer-reviewed journal articles provide an additional source of confirmation. The information taught in, and the building blocks of, a scientific field are consensus statements and deserve to be labeled credible scientific evidence. To this point, a definition of credible scientific evidence is fairly straightforward.

How, then, should statements from scientists that say explicitly or imply that "all science is tentative" be interpreted? If that is so, we must also conclude that credible scientific evidence is tentative. But how tentative is tentative? Most sciences have stable bedrock principles that change only if a major revolution occurs. Yet other scientific material is adjusted frequently. Approximately 10 years ago, my wife, Betty, persuaded me that we should become regular subscribers to, and readers of, health letters from the Harvard and Johns Hopkins University Medical Schools. I was not surprised that their recommendations changed, but I have been impressed by the amount of change that has occurred over this 10-year period.

What causes science to change? Philosopher-scientist Thomas Kuhn tells us two conditions must exist: 1) the accumulation of anomalies (failures in practice of the consensus view), and 2) the advancement of a different theory or concept. Each requires additional comment.

For a field such as medicine, there is great incentive to create improved treatments: change is rewarded financially. Anomalies (failures) are demonstrated in practice and recorded in scholarly journals. Yet in other fields of science, change may be less rapid, because workers in normal science may be defensive of their specialties (Kuhn noted this, and I have observed it in practice). It stems from the fact that some normal science scholars have a vested interest in the continuation of that in which they have expertise and where they have established reputations.

The new theory or concept must accommodate at least part of what the old theory or concept covers and must explain at least some of the anomalies or failures of the prevailing theory or concept. New discoveries and theories occur infrequently. The bottom line is that credible scientific evidence is a dynamic rather than a static concept. It does not, and cannot, play the same role in all public policy situations over time. Such considerations are well known by the scientific community; they are not common knowledge among all those who would use credible scientific evidence.

OPPORTUNITIES AND CHOICES

Imagine a coin with one side reflecting opportunities and the other choices. Suppose further the coin applies to the experiences of an applied economist in mid-career. The applied economist cannot know everything that was in the mind of those who had made opportunities available to him or her. Conversely, others cannot know everything the applied economist took into account when professional and personal choices and decisions were made at different times in his or her career.

Both opportunities and choices may be better understood, however, if the philosophical orientation of the applied economist can be known or deduced. First, consider opportunities. If a prospective employer or administrator is considering creating an opportunity for the applied economist, the employer may deduce a great deal about the applied economist from the past personal and professional choices and decisions he or she has made. Next, consider choices. The applied economist, consciously or unconsciously, is likely to make choices consistent with his or her philosophical orientation.

The above thought experiment has been useful to me as I have reflected on the opportunities that have been made available to me and the choices I have made in response to those opportunities. In the course of writing this book, I have become conscious of the number of relatively unstructured professional opportunities that have been made available to me. On the other side of the coin, much of the appeal of these potential assignments may have stemmed from the fact that it was unlikely they would be highly structured. In other words, the potential assignment was consistent with the way a pragmatic and pluralistic applied scientist would approach such a problem.

Additionally, one's philosophy will affect personal as well as professional choices. I know full well that both my personal and professional choices are affected by past experiences and study, both formal and informal. I am fiscally conservative with my own resources, as well as with the resources of others entrusted to me. No doubt that attitude was seared into my being by those Drought and Depression years of the 1930s. Even so, the great economists Frank Knight and Joseph Schumpeter taught me that risk taking is necessary if certain accomplishments are to occur. In that respect, I think of the decision to use part of the RFF endowment for the acquisition of real property in Washington, D. C.

Until 2007–08, I thought of myself as a liberal Democrat. I had been a registered Democrat since I was first able to vote, consistent with my populist background. I viewed the Democratic political party as more sympathetic to the needs of those less able to speak for themselves. The Iraq war led me to thoroughly reexamine my political affiliation. I was deeply troubled that my nation would undertake a preemptive war, and I held the Republican Party responsible. After much thought, I came to hold the Democratic Party equally responsible. I believe it did not meet its responsibility as a minority party by failing to force a national debate about preemptive wars as such. Its defense has been that it was lied to about weapons of mass destruction. The implication of this is that, had weapons of mass destruction existed, the preemptive war would have been justified. This is a position I cannot accept, and I eventually dropped my membership in the Democratic Party as a result. Some of my liberal friends thought I was intending to become a Republican. I never had any such intention, and I find it liberating to be independent.

Pragmatism is not necessarily devoid of idealism. I call attention to the words of Abraham Lincoln: "Now we are engaged in a great civil war *testing* whether that nation or any nation so conceived and so dedicated can long endure" (the emphasis is mine). Lincoln was not certain the nation could long endure, but he knew the radical ideas on which it was founded would be tested. In the case of the Iraq situation, I wanted a debate as to whether a preemptive war would be justified *even if* weapons of mass destruction existed. The requirement of pragmatism is that the empirical consequences of beliefs, and actions based on those beliefs, be accommodated. I am encouraged by domestic and foreign policies that are consistent with professed national ideals. We need not impose such ideals on others in order to be internally consistent ourselves.[2]

There is little doubt my populist background and my pragmatic philosophy made me receptive to the land grant university ideal. Clearly, the land grants originated in this tradition and addressed an important need at that time and for many years thereafter. As noted in Chapter 7, I believe the continued viability of the land grant ideal will depend on the identification of, and response to, emerging and contemporary need.

Previous chapters gave considerable attention to special assignments that have taken me away from the substantive subject matter in agricultural, resource, and rural economics. As noted, it is likely my philosophical orientation has affected how and why these opportunities came about and why I chose to exploit them. I close this final chapter by directing attention to farm management, resource economics, and rural economics, the substantive fields with which I am most closely identified. At the end of the discussion for each of these subjects, I identify a major issue that study of those subjects has helped me understand.

THREE FIELDS—FARM MANAGEMENT, RESOURCE ECONOMICS, AND RURAL ECONOMICS

FARM MANAGEMENT: Large, commercial farms in the United States produce most of the food and fiber coming to consumers through established marketing channels today. Yet smaller farms are growing in number in many places as well. These smaller farms are a heterogeneous collection. Some are part-time. Most benefit from off-farm income. Some sell their farm output to well-established markets; others market their output directly. Farmer's markets have been increasing. Some of these meet the requirements for organically produced food; others do not.

Can traditional farm management, which draws most of its theory from microeconomics, provide some understanding of these developments? I believe it can provide *some* but not *complete* understanding, at least not in the space available here.

Consider first the larger commercial farms that produce most of the output that flows through commercial channels. The larger commercial firms typically sell large quantities of relatively homogeneous products. In doing so, they operate consistent with principles set forth by Adam Smith more than two centuries ago. Standardized products permit specialization in production processes. Cost per unit often declines as output increases. And output will increase if warranted by demand. Demand, in turn, has grown with increases in population and improved income, both domestically and abroad.

Just because commercial farm output tends to be homogenous and in large quantities does not mean that these farms produce only one product. Some farms specialize greatly, but this is not necessarily the case. Some diversify by producing more products, some do so by operating in different locations, and some create products that the markets judge to be different by sorting and grading processes.

Commercial farms have produced large quantities of food economically, as measured by market standards. There are two major criticisms. One is that commercial farming has abused the natural resources on which it depends—that it has eroded soil and polluted groundwater, for examples. Another is that commercial technologies, such as pesticides and herbicides, may have negative effects on a healthful food supply. Such possible effects are recognized in economic theory as market failures. Both governmental regulation and environmental organizations have come into existence to deal with perceived market failures in the production and marketing of the food supply. An obvious consequence of perceived market failure of commercial farming has been the increased interest in organic food production.

I turn next to the increase in relative numbers of smaller farms. Clearly, there are many forces at work. Some are hobby farms. Some are the result of increased opportunities for off-farm work. Some are the result of consumer desire for alternatives to commercial food supplies. Traditional farm management and microeconomics provide some understanding of how the certified organic and not-certified organic food supply competition has developed.

When interest in organically produced food supplies first became common, discussion immediately arose as to what production practices would qualify food supplies to be sold as being "organically produced." So long as such practices were not specified or enforced, the choice among suppliers was necessarily subjective. Under such circumstances, price per unit varied considerably. These were not conditions where large-scale, highly commercial farming, as described earlier, would flourish. When a consensus emerged as to what would qualify as "organically produced," it was possible to establish a legal requirement if food was to be advertised as such. This made greater standardization possible, a characteristic of large scale, commercial production. The result has been an increase in the number of larger farms that are producing substantial quantities of more environmentally benign output.

I anticipate farmers markets will continue in popularity in many locations. They provide consumers with the opportunity to acquire not just organically produced food, but also food with other attributes as well. Betty and I often search for particular varieties of fruits and vegetables with a desired degree of freshness. We expect to pay more per ounce, or per pound, than we would at a supermarket. Improved quality is one reason farmer's markets are more successful when they serve more affluent neighborhoods. Such markets may also sell entertainment and status. Although direct marketing has grown in popularity, it accounts for less than 2 percent of farm-produced food products.

The principles at work here have application far beyond food production and farmer's markets. Consider the goods and services supplied by governments of various sizes. I venture to say that large, centralized units of government work best when they can provide a homogeneous, standardized product. The United States Social Security program provides an example. The number of people served is sufficiently large so that even those who are in some way exceptions can be anticipated and grouped with others who share the same exception. One blueprint can be developed not only for the main stream, but also for groupings of the exceptions. The difficulties of doing this increase rapidly as the complexity of the services to be provided grows. The fate of centrally controlled economies around the globe during the past century provides support for the conjectures made here. For example, both the Soviet Union and the People's Republic of China drastically decentralized their respective highly centralized economies during the 20th century. In both cases, decentralization resulted in substantial increases in the per-capita production and consumption of goods and services.

My interest in size of decision unit, whether public or private, in public policy has arisen directly from my study of farm management and agricultural policy (see Chapter 7). There is a great deal of symmetry between public and private organizations as they increase in size. In the private sector, large organizations typically derive their advantage over smaller firms because of economies of scale with a standardized product. Public-sector large firms also perform best when their service can be standardized. Larger firms need not have great advantage over smaller firms per unit of output in order to achieve great profit if volume of output is sufficiently large. But if large firms lose their per-unit advantage for any reason, they may quickly face major deficit problems unless their reserves are great.

As recent experience has demonstrated, the failure of a large firm may inflict cost on others in society — "too big to let fail." Economic public policy literature generally does not consider size of firm to be of concern so long as competition is maintained. Yet recent experience demonstrates that large firms often do not internalize all of the risk associated with increases in size. One hopes that public policy literature will become more concerned with risk–size relationships as well as who bears the cost of risks associated with size. The recent experiences of Fannie Mae and Freddie Mac make clear that risk–size relationships are important in both public and private sectors.

RESOURCE ECONOMICS: When I went to Resources for the Future in 1975, a firm nonadvocacy, nonpartisan policy existed. I had no difficulty wearing this mantle and continued the policy after I became president. Given the nature of our subject matter and our mission, it was inevitable that we involve ourselves in some of the more controversial issues of our time. It is necessary to ask if we brought a point of view to bear as we investigated these issues. Near the end of Chapter 5, I noted that Irving Fox, former RFF vice president, maintained there was an "RFF point of view." According to him, that point of view was manifested in an allegiance to the economic efficiency model inherent in microeconomic theory. This must be viewed as a generalization with exceptions. Not all RFF researchers thought or believed the same. Even so, when I was there to make direct observations, I believe Irving's Fox's generalization still held. This point of view reflected the belief that economic efficiency is important in our society. Furthermore, the economic efficiency model as set forth in economic theory provides a useful way of evaluating public policies.

I mention the RFF experience here only to underline the practical importance of the discussion to follow. As this is written, green public policies are being advocated widely. How are they to be evaluated, if at all? If they are to be evaluated, by what standard will they be evaluated, and who will conduct the evaluation? Such questions are of greater importance than many technical discussions that command space in resource and environmental economic journals.

Knowledge about the economic efficiency of policy options is useful because it is the means by which more ultimate ends can be achieved. Both pragmatism and pluralism raise caution flags, however, about making sharp distinctions between means and ends. What constitutes an end to some may be a means to others. It may be more accurate to think of a hierarchy of ends rather than categorizing desirable outcomes as either means or ends. In either classification, economic efficiency gains are likely to be viewed, not as an ultimate end, but rather as a way to facilitate the achievement of other ends. During the Drought and Depression years of the 1930s, a "steady job" would have been described by many as a near-ultimate end. One of the most envied people in my community was the rural mail carrier. His was a steady job with additional benefits. If his envied situation had been analyzed carefully, I am confident it would have been discovered that the reason a steady job was desired so greatly was that other, more desirable, outcomes could not be obtained in the absence of a reliable income or wealth. It was a necessary condition for achieving more ultimate ends as, for example, family medical services or education for a gifted child. I consider improvements in a nation's or a community's economic efficiency to play a similar role in public policy analysis.

The analyst may believe that when the economic efficiency effect of a given policy has been estimated (yes, estimated), his or her task has been completed. The story should not end there, but often it does. Members of the body politic may not believe that which would be sacrificed (say x acres of wilderness) to achieve some level of economic efficiency is worth it. In other words, estimating the economic efficiency effects of a proposed policy change may inform the policy process but is not a substitute for it. As an example, the Endangered Species Act is in the nature of a categorical imperative. This Act does not stipulate that economic considerations are totally irrelevant to species preservation decisions, but economic consequences are not permitted to trump species preservation objectives under the strictures of this Act. In other words, economic efficiency improvements are not an ultimate end in public policies.

Both economic and ecological predictions are typically clothed in uncertainties. This often results in the predicted outcomes of proposed policies themselves becoming subjects of controversy. Under such circumstances, projections and predictions of policy analysis may play a small role in policy outcomes. My friend Bill Wilkins writes: "Full employment will do more for a nation than soulless micro efficiency."

MY BOTTOM LINE: Economic analysts should not be expected to find that green investments will always yield handsome economic returns. Neither should economic analysts expect their economic efficiency findings to automatically become public policy. To state the matter more bluntly, potential economic gain is important, but only if it will permit something even more important to be obtained. To decide, it is helpful to know what potential economic gain is feasible and what it can achieve if it is realized.

In recent years, considerable attention has been given to natural resource management that will "meet the needs of the present without compromising the ability of future generations to meet their own needs" (World Commission on Environment and Development 1987). "Sustainability" has become a focal point of much resource and environmental economics literature. This literature contrasts the sustainability objective with more traditional economic objectives. The sustainability objective reflects a value judgment that establishes a relationship between the "needs" of present and future generations. Clearly, resource allocations are likely to be different in the two cases, with the sustainability case generally regarded as being more "green." Arrow et al. conclude that global economic performance has not been optimal when measured by the sustainability objective. This is associated with the under-pricing of certain natural resources.

When the two objectives are compared from the viewpoint of central planning or central control, certain parameters do not change. The economy and ecology remain important in both cases. Further, similar uncertainties are associated with assumptions about, or predictions of, the future condition of each system and the performance of both. The central planner who wishes to contrast the probable future performance of two systems (one sustainable, one economic) will start with different constraints placed on each (say in T_1). From T_1 forward, the adjustment problems facing each will be identical in many respects. Both will need to contend with changes in the economy, the natural environment, human knowledge (technology), and perceived "needs" of existing and future generations. Furthermore, the central planner must consider the adaptations each system makes to new information as it moves through time. Adjustments will not be identical because the sustainability constraint will make some difference. Even so, how complex systems generate feedback information and respond to new information may swamp initial constraints and original starting points. The evaluation of complex systems may be different if viewed ex ante than if viewed ex post. A "green" system viewed ex ante may turn out to be "brown" viewed ex post if the system is not an adaptable one.

The above discussion directs attention to how decisions get made as systems move through time. What decisions are made from a central point and what decisions are decentralized? Do decisions made at a central point influence incentives for decentralized decisions? How are the institutions designed that provide the legal framework for all decisions, both centralized and decentralized? Does the institutional framework generate and provide feedback of information that will influence future performance? Such questions are central to long-run sustainability. Too often, sustainability is interpreted to emphasize stability and the stationary state. Continuous adaptability and adjustment likely will be of equal or greater importance.

In Chapter 7, there was discussion concerning the relationship between energy and the environment in economic growth. The relation of energy and the environment to economic growth, and the intergenerational issues arising from the sustainability movement, demonstrate the fundamental nature of resources and the environment in society.

RURAL STUDIES: Shortly after I went to Oregon State in 1954, I became aware that my chosen field of agricultural economics was neglecting many of the messy problems affecting rural America. My reaction then and for many more years was, "that is not for me." Despite my pragmatic orientation, I preferred the precise problem formulation that was possible in farm management and resource economics. When the time came for me to leave RFF, but I was not yet ready to retire, I decided to find out if I could make a contribution in rural studies.

To my considerable surprise, I was able to obtain a major grant from the W. K. Kellogg Foundation to establish the National Rural Studies Committee soon after I returned to Oregon State in 1986. The formation of that group and its accomplishments are described in Chapter 6. It was soon clear why progress was difficult in rural studies. The field lacked a central core, and the best scholars were scattered across disciplines and institutions. Chapters 6 and 7 present details on these problems as well as recommendations and proposals that address some of these difficulties. The process of writing this book has resulted in my attention being directed to the subject again. Original thoughts on the topic can be found in Chapter 6, in the section that describes the National Rural Studies Committee assignment.

Chapter 6 also proposes the establishment of a multi-disciplinary group to advance rural studies. That is a worthy goal, and the rural place theory set forth in Chapter 6 is of importance as well. The view set forth there stipulates that rural America is heterogeneous, dynamic, and turbulent—characteristics not captured adequately by existing theories of the location of economic activity. It suggests the need for, and provides material useful in the construction of, a new spatial economics. This new spatial economics would include central place theory, based on a model first advanced by von Thunen in 1826, which constitutes the heart of regional economics. Urban economics rests on that model, supplemented by agglomeration economic concepts and insights from the new growth theory. A literature has also developed about the formation of edge, or galactic, cities that generate their own separate dynamics as metropolitan places grow (Lewis 2005). The new spatial economics would integrate and accommodate these separate views of different parts of the economic landscape. This would permit identification and measurement of linkages and interdependencies among the hinterlands, the cities, and the places between. It would also provide a framework for considering the adequacy of existing property rights systems and other institutions to address problems arising from the simultaneous operation of both centralizing and decentralizing forces in rural places.

CONCLUSIONS AND CONJECTURES

In drawing this book to a close, I attempt to assess the philosophy I hold with respect to fundamental questions that have arisen from a professional lifetime of intellectual effort in agricultural, resource, and rural economics. Examples of such fundamental questions include:

- What are the returns, and who bears the risks, of public and private organizations of varying size in society?

- What is the relation, both direct and indirect, of energy use in economic development to the natural environment? Are there limits to human ingenuity to compensate for the modification of nature as economic development occurs?

- What are the necessary features of a system of property rights in rural places that will accommodate the commercial production of commodities, enjoyment of natural amenities, and disposal of the residuals generated by both urban and rural economies?

The philosophy I bring to the investigation of such questions can be described as pluralistic and pragmatic, informed by the experience of society and illuminated by the conceptual insights of humans. In other words, pragmatism makes use of theory, and pluralism means that more than one theory will be considered. Pragmatism requires that the truth of propositions is judged by the test, "does it work?" This, of course, is not a simple question. The answer will depend on experience, with experiments being a possible part of experience. And, for many questions, "learning by doing" becomes an important part of experience as it helps decide what it is that "works" or "does not work." It is my judgment that economic history is a neglected part of research methodology in economics, and I suspect the same is true of research in other fields as well.

I offer comments next about economic theory and its contribution to the pragmatic question "does it work?" Before offering my assessment, I wish to say that I truly love and admire much economic theory. It has been a significant source of intellectual stimulation and conceptual beauty in my life. Nevertheless, I view the practical contribution of economic theory differently than do many economists. In brief, I consider economic theory to be *my servant, not my master*. For many years, I believed it was only recent graduate students who thought otherwise. Graduate students in economics often struggle greatly for two, three, or four years to master and relate various parts of economic theory. It is understandable if some come to view economic theory as reality, rather than an abstraction of reality. If economic theory were reality, economic public policies could be deduced directly; there would be no need to ask "will it work?"

The way economists view their theories is of practical importance because many economic questions end up in the public policy arena in one way or another. If economic theory is the master, rather than the servant, recommendations from economists likely will be stated and expect to be considered as "settled." Non-economists, then, are confused if contrary positions are set forth by other economists, also as "settled." All theory is, in some sense, of a partial nature. All of reality can never be captured by a conceptual model. Economic models may be useful points of departure when complicated situations are confronted. I try to avoid the seductive notion that the presuppositions and assumptions implicit in economic models are necessarily realistic. As noted in Chapter 7, we may not know what we may not know.

Recent events can serve to illustrate the message of the previous paragraph. When the economic crisis of 2008 was at its worst, popular disagreement among economists revolved around the optimal amount of regulation of capital markets. Non-economists learned from these disagreements that Freshwater (University of Chicago) economists favored decentralized markets with minimal regulation, and that Saltwater (Harvard, MIT, Berkeley) economists were more inclined to favor government intervention. Saltwater economists implied that the recession was the fault of the Freshwater economists, because proper regulation would have prevented excessive speculation and bubbles. Freshwater economists expressed fear that recovery would not occur, or be delayed, because excessive government regulation would prevent the proper functioning of markets. Neither group had predicted the crisis. Nor had either group called attention during the good times to the failures of Enron and Long Term Capital Management as harbingers of bad things that might occur.

Near the end of his life, and after the crisis had developed, the great economist Paul Samuelson spoke in a television interview of the failure of economists to predict current economic difficulties. He then said that economists should accept some of the responsibility for current problems. He specifically mentioned the lack of regulation of new financial instruments that had been invented by economists associated with his own institution (MIT). These instruments had also played a role in the earlier Long Term Capital Management failure. It is my view that the various economic theories relied upon by Freshwater, Saltwater, and other economists are partial views of the "real" economy. *All are useful for particular purposes; none is sufficiently robust to be our master.*[4]

Multiple theories, or economic constructs, can be a source of strength in democracy if opposing views are brought into the open and stimulate debate. During my professional lifetime, I believe the major public policy mistakes that have been made can be traced to some form of "group think" when policy was established, or when there was a refusal to reconsider existing policy. Science is not immune to this danger, resting, as it does at any given time, on some form of consensus. Yet with science there are built-in practices that, if followed, will result in change. That is why data availability, peer review, and transparency generally should be taken seriously. There are dangers in science if a particular scientific paradigm becomes closely affiliated with a particular political position or specific financial interests. "Political correctness" and lack of tolerance for dissent in academic life is but one example of "group think" tyranny.

No doubt my concern about "group think" underlies the feelings of liberation I experienced when I no longer was a member of any political party. I have long known group action, based on consensus, may be more effective in achieving individual ends than individual action independently determined. Yet I am reluctant to make open-ended commitments that may require support for unanticipated causes or positions. In principle and practice, I have no objection to government intervention in markets in particular, or in many private-sector activities generally. Nevertheless, for matters of fundamental importance, I would rather depend on the judgment on many independent decisions than I would the consensus judgment of elite groups. In conversation with elite intellectual establishment types, I often have been shocked by their naïve stereotyped views of "unwashed groups" in our society. Early in my intellectual journey, a philosopher friend often said: "No person is wise enough, or good enough, to have dictatorial power over any other."

Were I to undertake a second intellectual journey, I would certainly choose a route that would permit me to apply and make use of intellectual constructs. The opportunity I have had to explore the conceptual and to help with the practical is perhaps the greatest privilege I could have had. Those family, friends, students, and colleagues who accompanied me on my journey seemed, while not having identical interests, to understand why I needed space to find myself and to be myself. Were I to travel this road again, I would retain my pluralistic and pragmatic orientation, but I probably would elect to be even more independent. Even though it is lonely at times, independence has its virtues. It does not require a choice between "left" or "right" along a continuum; rather, it liberates and empowers a search for new dimensions altogether.

Appendix A

THE VICTIM AND THE CAREGIVER:
COPING WITH DEMENTIA
A True Story: 1987–1999

Merab, my partner in marriage for more than 53 years, died August 4, 1999. She was diagnosed as having Alzheimer's disease in 1987. Thus, we coped with dementia for at least 12 of the 53 years we were married. I write "at least" deliberately, because she suffered with the disease for an unknown period before the diagnosis was rendered. With Alzheimer's, the victim and the caregiver travel together over unknown and unpredictable terrain. Alzheimer's is a journey, not an event, and it wends a circuitous route. There were a number of people who made the journey much less painful than it otherwise would have been. Some of these people are identified in the course of telling this story. In particular, the love and support of our daughter, Cheryl, and her husband, Bob, contributed immensely.

There was a fundamental change in Merab's well-being at least two years prior to the diagnosis of Alzheimer's. We were living near McLean, Virginia, in 1984 when Merab began to show signs of the depression that had afflicted her from time to time throughout her life. This time it was different, because the depression was associated with a loss of cognitive ability. This was disturbing to me, and it must have been terrible for her. That Merab might be suffering from some type of chronic dementia was not suggested by her physician, nor did it occur to me.

At that time, I was in my early sixties. I was president of Resources for the Future, a public policy research institute (think tank) located in Washington, D.C. I was in good health and enjoying my work. Nevertheless, I became convinced Merab's condition required that we relocate to where we were going to spend our retirement. Resigning and relocating was not a simple matter. I needed to give my organization sufficient notice so that a successor could be chosen. We had lived in Corvallis, Oregon, home of Oregon State University, where I was employed prior to moving to McLean in 1976. We decided to return to Corvallis and did so in March 1986, and there Merab spent the remainder of her days.

Each phase of Alzheimer's has its own distinct characteristics. Both the victim and the caregiver have different demands placed upon them, as well as their relation to one another, as the disease progresses. The prediagnosis period may be of short duration and not of great consequence for some. With others, it is stressful indeed.

As Alzheimer's progresses, it deprives the victim of his or her ability to perform functions that were once routine. It is never easy to relinquish those things we enjoy and give us status or satisfaction. In the prediagnosis phase, the process of surrendering functions has not yet been legitimized medically. The caregiver, who may not yet even know he or she *is* a caregiver, may resent the victim's behavior because their own lifestyle is being affected. Both the victim and the caregiver may hold the victim to pre-dementia standards. They typically cope alone with what arises because they may not know the kind of outside assistance they need.

EARLY ALZHEIMER'S

After we returned to Corvallis in the spring of 1987, Merab underwent a review of her medical condition. Her primary care physician arrived at a diagnosis of Alzheimer's based on the elimination of other possible explanations of her symptoms.

The diagnosis changed my frame of reference significantly. I had to accept the reality that Merab was never going to improve markedly.

Indeed, I could only expect a continuous, irreversible decline. The person I had known over the years was no longer with me, and would never again be with me, even though personality characteristics remained until the end. My daughter, Cheryl, and I recognized we were dealing with a problem not capable of being solved, but rather, a situation we needed to manage. Perhaps we were fortunate we did not know the situation we were to manage was to continue for more than 12 years.

The distress of the caregiver at this point should be placed in perspective relative to the waves of terror that must be washing over the victim. The victim no doubt senses, however dimly, that control of the nervous system and body is diminishing. And he or she may sense the status once enjoyed in the eyes of others is changing. Some become depressed, some withdrawn, and some engage in aggressive or outrageous behavior. Most, at some point, become highly suspicious of those around them. Caregivers often are badly hurt when they discover the person for whom they are sacrificing is questioning their motives.

Early on, I attended several Alzheimer's support groups. Sharing experiences was much more valuable than having someone tell me how to cope with a particular problem. Also, I recognized I was more fortunate than most of the support caregivers I came to know. First, I had enough income to obtain help for some of Merab's care; many caregivers did not. Second, Merab's personality was such that in the early stages of her disease, she was more inclined to withdraw than to engage in aggressive activities. Third, my professional work provided for renewal from the stress of caregiving. My greater fortune led me to recognize that the victim and the caregiver are linked in such a way that the welfare of one cannot be judged independent of the welfare of the other.

Alzheimer's almost always results in significant costs being borne by someone in society. Who bears those costs depends on the way the disease develops, how well the caregiver handles difficult situations, and the financial status of the parties involved. The health of the caregiver may not permit her or him to take on all the added duties, and, indeed, there are many. Traditionally, neither medical insurance nor long-term care insurance has been of much help, at least in the early stages of the disease. Medical insurance usually will pay only if skilled nursing is required. Long-term care insurance is expensive and typically cannot be drawn upon except in the more advanced stages of the disease. In Merab's case, I cared for her in our home, with hourly help, for more than six years, and she was in a nursing home for a comparable period. This meant the cost of her home care and her nursing home stay was our responsibility. These costs were in excess of $400,000 over the course of the disease. Covering costs of this magnitude simply is not an option for many people. Some declare bankruptcy and obtain help from Medicaid. Others just "do it yourself" with all the sacrifices that entails. Merab and I had savings that permitted me to have help when she was at home and for her to be in a nursing home of the highest quality when that became necessary. I am indeed grateful our modest holdings could be used in this way during our time of need.

As noted, Merab had a tendency to withdraw from social contact as well as physical activity. I attempted to keep her involved as long as it was possible to do so. Many household tasks were soon beyond her ability to perform them. We went together on numerous trips. Some were made in connection with my professional work. I grow and exhibit roses, and Merab attended many rose shows with me. Cheryl and her husband, Bob, arranged for us to cruise with them on the inland passage to Alaska. Travel was always a challenge, but such activities helped Merab maintain social contact far longer than she otherwise would have.

When I returned to Corvallis in 1986, I accepted a part-time position at Oregon State. When Merab was diagnosed as having Alzheimer's, my immediate reaction was that I should bring my professional work to a close and devote my time to her care. My friends, relatives, and colleagues all thought that was a bad idea. Rather than make a decision that might be difficult to reverse, I "punted" and simply postponed severing professional ties. I then came to realize that I would have made a major mistake if I had done otherwise. Continuing professional work was good for me, it was good economics, and Merab was better off as well because I could do more for her when I could renew myself from time to time.

Merab's condition worsened with the passage of months and years. It was difficult to coordinate my sleeping pattern with hers when she was at home. Often she would sleep for extended periods, perhaps arising

Merab Weber Castle   Cheryl Diana Castle Delozier

only for meals. Those sleeping periods usually would be followed by long stretches when she would not sleep at all. As time passed, I did not want to sleep unless she was also asleep. Although I did not realize it at the time, I was afflicted with sleep apnea, and I suffered increasingly from sleep deprivation during the six years I cared for Merab in our home.

During this period, I developed a "childhood in reverse" concept to help me better rationalize Merab's slow decline. Parents have long relied on authorities to tell them of developments they can expect in their children at different ages. It occurred to me the reverse existed with Alzheimer's. Even though the "childhood in reverse" concept did not enable me to predict precise reversals, I was better able to accept them when they did occur.

Alzheimer's victims often have multiple caregivers during the course of the disease. Heart attacks, strokes, and nervous breakdowns send some caregivers to the sidelines temporarily or permanently. Outsiders can never know of the full range of pressures that exist in particular circumstances. I have resolved that I shall never pass judgment on the behavior of a caregiver because I can never know the totality of the situation with which they must deal.

THE ACCIDENT

Merab was still in our home in the summer of 1993. She had become incontinent, and her sleeping pattern was more erratic. Although I did not know the full extent of my own difficulties, I sensed I was becoming very fatigued. About 18 months earlier, I had asked Cheryl and Bob to investigate the nearby Mennonite Home, a nursing facility in Albany, about 15 miles from our home in Corvallis. Cheryl and Bob were favorably impressed, and I visited there. I was advised to place my name on their waiting list even though we were not yet ready to use their services. The Home then called me at 6-month intervals to ask if I wished for my name to remain on the waiting list. Thank goodness I always renewed.

One Sunday evening in the summer of 1993, Merab was quite restless. I helped her make preparations to retire at her usual time between 9:30 and 10 p.m. She did not go to sleep right away, and I gave her a mild sedative that had been prescribed by her physician. She was still awake at 11:30, and I was becoming very tired. Our physician had told me I could give two sedatives if it were necessary. That night I gave her the second sedative and made my own preparation to retire. I went to bed at midnight and promptly fell asleep, confident she would fall asleep as she always had. At 2:10 a.m., I awakened to find she had left her bed. I looked inside the house, checked outside, and when I did not find her, called 911.

The Corvallis Police Department responded immediately. They asked me to remain at home. The officer in charge came to our house and used it as his headquarters as he directed the search. The police canvassed the neighborhood with patrol cars, decided Merab was within a relatively small area around our home, and then conducted the remainder of the search on foot. Less than two hours after I reported her missing, she had been found. When discovered, she was lying in tall grass on a vacant field about three long blocks from our home. Apparently she had started down an incline and then fallen. Her fall destroyed a great deal of the bone in her left wrist and nearly severed an artery. An ambulance was summoned, and she was taken to the hospital emergency room. I met her there approximately two hours after I made the 911 call.

Merab's injury required surgery later that day and then again 10 days later. After surgery on the first day, she was taken to a nursing home in Corvallis where she remained until her second surgery. When the medical staff understood the extent of her injury and the kind of surgery required, they said home care was out of the question, at least for a time. When contacted, the Mennonite Home said I would have the right of first refusal on a bed because I had been on the waiting list for 18 months. Merab was taken there from the hospital the morning after her second surgery.

Merab lived the final six-plus years of her life in the Mennonite Home of Albany. I did not have a single serious complaint about her care during that period. At the time of her accident, I believed the care Merab received in our home was good to very good. In retrospect, I continue to believe that is a fair evaluation. Within a week after she went to the Mennonite Home, I recognized she was far better off there than she was in my care. I never left those premises concerned about Merab's care while I would be away. I knew those competent people would tend Merab as they would a loved one in their own family.

I do not know what would have happened to us if Merab had not had her accident. I doubt I ever could have brought myself to make the decision that would have her taken from our home. Did I have a moral right to have her leave a place that was hers as much as mine? After she was admitted to the Home, and I observed the care she was receiving, I had no guilt feelings about her being there. If the Home had been unsatisfactory, I am not sure how I would have reacted. How one handles one's guilt is an issue with which all caregivers must deal.

TURNING POINTS

Because the victim and the caregiver are so closely linked, it is important to identify when the relationship changes in a meaningful way. Such a change usually will be triggered by a change in the victim's condition, but the picture will not be complete unless the effect on the caregiver is considered. Three significant turning points can be identified during Merab's illness. All involved the needs of the victim and the response of the caregiver.

The diagnosis of Alzheimer's marked the first turning point. The diagnosis legitimized her illness, made clear the need for a caregiver, and established the boundaries of our new relationship. The acceptance of this reality was a profound emotional experience. I never doubted I was going to help Merab as long as I was able, nor did I believe life had dealt me an unfair blow. But I was sad that I needed to say goodbye to a relationship of many years, and the major uncertainties that loomed were frightening indeed.

The second turning point came when Merab left our home after her accident. We had shared the same residence for nearly a half century. Her care had been a major burden, but her departure created a huge emotional void. Our relationship was changed again in a fundamental way. She continued to need me, but her needs had changed, and if I was to help her, it was necessary I adjust accordingly.

The final turning point came with Merab's physical death. Again, a major emotional experience, but this time there was much support from extended family, friends, our church, and others. After the physical death, the caregiver can work to bring closure to the trauma of having a loved one suffer at great length from chronic dementia.

The turning points that existed for Merab and me will not be the same for all. Nevertheless, turning points in particular cases need to be identified and understood. My experience might have been less painful if I had been sensitive along the way to turning points in the victim-caregiver relationship.

BETTY

As noted, Merab went to the Mennonite Home in the summer of 1993. She was always on my mind in one way or another, but her absence from our home resulted in a host of my emotional needs coming to the surface. For example, I longed for stimulating, intelligent conversations generally, but especially with members of the opposite sex.

By the spring of 1994, I resolved to develop a social life appropriate to my situation, but I did not know what that should or could be. The opportunities are not great for a married guy in his seventies who is without a spouse. When the roses came into bloom, I held a modest open house to which I invited a few neighbors, colleagues, and members of my church. Betty was one of the people who attended. Betty and her spouse had moved to our neighborhood to retire shortly after

Merab and I returned to Corvallis in 1986. Betty's spouse developed brain cancer not long thereafter; she had been a widow four years when she came to my house in 1994. She had befriended Merab shortly after we all came to Corvallis by taking Merab to some church functions.

Betty Rose Burchfield-Castle

After Betty attended my open house, we chatted from time to time in church and in grocery stores. I finally summoned the courage to call and invite her to my house for a dinner for two. She accepted, and we discovered we had a great deal in common. She was reared in South Dakota; I was reared in Kansas. She is the third child in a family of six; I am the third child in a family of four. She had three children and, at that time, four (now five) grandchildren.

Early in our developing relationship, we discussed frankly the circumstances facing us. She said if we were to go forward I must "never neglect Merab." I responded that to do otherwise would destroy the foundation of any relationship that might develop. We never again discussed the subject.

We neither concealed nor flaunted our companionship. We did not ask our children for approval, but they were the first people we told of our regard for each other. We agreed we could not be concerned by the opinions of others. It may have been more difficult for Cheryl to accommodate the thought that there was a person in my life other than her mother than it was for Betty's children to accept me. Merab was still living; Betty's spouse had been dead for some time. When Cheryl came to know Betty and understood what she was doing for me, they developed great mutual respect. Betty and I soon developed a pattern of having our evening meal together. We went together to social events, including church attendance. We have taken trips, but as long as Merab lived, we were never away from Corvallis for more than a few days at a time. Betty's companionship improved the quality of my life immensely.

Neither of us knew how our relationship would be viewed generally. Any concerns we may have had on that score were dispelled by the attitudes displayed. Betty is respected highly by all who know her and deserves credit, I am sure, for much of the acceptance we have experienced. She displayed great judgment and tact when she faced potentially awkward situations. During the days just before Merab's death, Cheryl, Bob, myself, and a friend of Cheryl's, Audrey Snuggs, were at Merab's bedside as much as possible.[1] Each day, Betty brought our evening meal to my house when we returned from the Mennonite Home. She was in the background, but she contributed greatly after Merab's death until my family left town after Merab's memorial service.

Neither Betty nor I feel a need to justify our relationship. Nor do I wish to suggest the kind of relationship Betty and I developed necessarily will apply to other situations. Rather, the intent is to call attention to the emotional needs of the caregiver when the victim suffers an extended illness. This is a neglected subject, especially when the victim and the caregiver are partners, and deserves to be discussed in a forthright way.

THE FINAL PHASE

The caregivers at the Mennonite Home came to understand Merab right away. They realized she had taken pride in her appearance and they did as well. Her hair usually was well attended, and simple jewelry enhanced her appearance. The caregivers let her know they knew she had been a teacher early in her life. Merab adjusted rapidly to life at the Home. When she first went there, she was more social than she had been for some months and spoke to everyone she met on the premises. At times she thought she was in a school of some kind but was uncertain as to just what role she should play in such a setting.

One morning, she wheeled her chair past an alcove where three administrators were talking and drinking coffee. When she passed the alcove sometime later from the opposite direction, the three were still there. Shortly thereafter, she returned to the alcove where the three remained engrossed in conversation. She then propelled her wheelchair into the alcove, pointed to each administrator in turn, and said, "Detention! Detention! Detention!" When I went to the Home later that day, three different workers stopped me on my way to Merab's room, asking, "Have you heard what Merab did today?" The incident has become a part of the folklore and oral history of the Home, but is appreciated more by rank-and-file workers than by administrators.

Merab's "sleepy days" gradually became more frequent as her time in the Home lengthened. After a time, she became more serene. When she was awake and alert, she seemed to be at peace with herself and her surroundings. From time to time, she would have infections but would defeat them, although her vitality was affected with each bout. The nurses marveled at the strength of her immune system. Her ability to speak declined rapidly during the last two years of her life. She seldom uttered an understandable word but often comprehended what was said to her. One experience confirmed that was, indeed, the case. She suffered from polymyalgia and often would cry out in pain when she was touched or moved without warning. Her principal nurse designed a simple experiment. The experiment revealed that Merab would cry with pain if she were moved in her bed without warning but, if told, would seldom vocalize pain. Apparently information reached her brain, but little or nothing could move in the opposite direction. It is not surprising that Alzheimer's victims typically become exceedingly frustrated at some stage of their disease.

As the end approached, more and more decisions had to be made. During the course of Merab's disease, we requested that heroic measures not be exercised. We specifically directed she not be given antibodies to fight the numerous infections she developed. The issue was not how to keep her alive, it was how to make her more comfortable. When it became apparent she did not have long to live, we permitted her to remain in her bed. She never had bedsores or developed other problems that often afflict those who are bedridden. She passed away about 7 a.m., August 4, 1999. The cause of her death officially reads: "Bronchopneumonia due to, or as a consequence of, Alzheimer's disease."

Her memorial service was held August 6, 1999. The program for the service carries the following words over Cheryl's name:

It is not necessary to use many words to describe my Mother. To say she was a *lady* and the *essence of love* summarizes the person she was throughout her life.

To know her was to love her and to be loved. Many thanks to the Mennonite Home for recognizing the great woman she was and, despite her illness, helping maintain her dignity until the end.

Merab—*lady* and *love* were synonymous.

The hymn closing her memorial service carries the title, "Morning Has Broken."

Appendix B

A THING OF BEAUTY:
A ROSE IS A ROSE IS A ROSE
1955 to the Present

This is about my sole hobby—growing roses, sharing roses with family and friends, exhibiting roses, and helping tend public rose gardens. This is an appendix because, so far as I have been able to determine, my hobby has no connection with my intellectual journey except, possibly, as an escape from intellectual activity. Even so, I am known to numerous people mainly or solely because of my hobby. Some people believe that is all I know anything about; they would think it strange should they happen to see a book of this kind and roses were not mentioned.

ROSES ARE FUN

We moved into our new house in Corvallis in April 1955. Shortly after that my colleague D. Curtis Mumford, famous rose enthusiast, came into my office. As I recall the conversation, it went like this:

CURTIS: "Will you have roses in the garden of your new house?"
EMERY: "Curtis, my wife and I have been married nine years and she has moved with me eight times without complaint as I have acquired an education and become established. We hardly

have enough savings to put a carpet on the floor. I believe I should give attention to doing things inside our house before I spend money for landscaping."

CURTIS: "I understand." And he departed.

Several days later, Curtis returned and said: "I have found a nursery going out of business and they have some 'out of patent' rose bushes they wish to sell. Would 35 cents a bush be too much?"

During the 1955 Christmas holidays, I planted 13 rose bushes in the mud east of our new house. The thirteen comprised 12 hybrid teas and one hybrid perpetual. Hybrid teas are the most popular rose grown. Hybrid teas are the result of a cross between hybrid perpetual and tea roses. The hybrid perpetual provided for season-long blooming. The tea rose contribution to the cross was an increased probability that blooms would occur singly at the end of a long stem. The first hybrid tea is generally described as having been created in 1867, although there is controversy about that date.

The Corvallis Rose Society (CRS) was established in 1955 by the Men's Garden Club of Corvallis and was given the responsibility of holding an annual rose show. The Men's Garden Club had been in existence several years and had held an annual rose show prior to the creation of the CRS. I joined the CRS in 1956 and, at the urging of Curtis, entered my first show in 1957 (I believe). I won the novice trophy and two years later won the sweepstakes trophy for best in show. I was "hooked."

Curtis then began to educate me about other opportunities to exhibit. He told me about the Salem rose show and then the Portland rose shows, one in the spring and one in the fall. Curtis said the Portland Rose Society was one of the largest in the nation and that competing in the Portland show was analogous to playing in the "big leagues" in baseball. After a few years, I developed sufficient confidence to enter in Portland.

As I considered becoming an exhibitor, I became aware of the dark side of rose exhibition. I observed intense disagreements within husband-and-wife teams as they made rose show entries. Rose exhibition can be a stressful activity, especially if the show is a distance from home.

L—Double Delight at its most beautiful stage; R—Double Delight, Queen of Fall Rose Show, Portland, Ore.

Entries usually must be made by 9:30 or 10 a.m. if the judging is to be completed by the time the show is scheduled to open for the public, usually not later than 1 p.m. I also observed that some exhibitors do not behave gracefully when their roses do not win. Some question the ability and judgment of rose-show judges. And then I made an important discovery—most rose-show judges have poor eyesight! Clearly it would be sad if judges were to be criticized for decisions or judgments resulting from a condition over which they have no control.

I concluded that exhibiting should be undertaken only if it were going to be *fun*. It would not be *fun* if it caused family discord. And it would not be *fun* if one were to become angry with judges. I hope my family has not suffered unduly from my exhibiting roses, although I know sleep has been lost on some rose-show days. My family has been tolerant and supportive, and a family member often has accompanied me and helped enter roses. And I have neither quarreled with nor criticized judges. I did decide to exhibit, and it has been *fun*.

When we made our home in McLean, Virginia, during the RFF years, we joined two rose societies—the Arlington Rose Foundation and the Potomac Rose Society. I enjoyed both groups immensely and learned a great deal about rose culture. I exhibited only in two local shows during the years we lived in McLean.

There have been some tangible accomplishments associated with my hobby. I have served three times as president of the Corvallis Rose Society. I am accredited by the American Rose Society as a rose judge. I was selected in 2007 as the outstanding rose judge in the Pacific Northwest District of the American Rose Society. I have won Queen of Show approximately 25 times in shows at Corvallis, Salem, Portland (twice), Eugene, Tri-Cities, Albany, and Coos Bay. My greatest accomplishment as an exhibitor, perhaps, was to win the best 12 roses of a kind (called a collection) in the Portland spring show four consecutive years, each time with a different variety. I have been fortunate to have been a member of the two largest rose societies in the United States—the Potomac and Portland societies.

BEAUTY CREATION

I have reflected as to why it is that roses are an attraction for me. I believe it is because roses are a way to create beauty. I love music but I cannot carry a tune, and I cannot make music in any other way. I have no discovered talent in the visual arts, and no such latent talent appears to lurk beneath the surface. Originally, I was attracted to exhibiting roses because I enjoyed the competition. After a while, however, I began to think of rose shows in a different way—that is, as a way to discover and display beauty. I consider the best exhibitor I know, Dennis Kosmo, to be a seeker and creator of beauty. As he grooms a rose for entry, he radiates a driving passion for perfection. I do not believe he cares at all about who, or what, he will compete against. I do believe his motivation is to make his entry as beautiful as he can possibly make it.

This then raises questions: What is beauty? How do we know what is, and what is not, beautiful? The American Rose Society has prepared a handbook for judging roses that provides rules for both exhibitors and judges. The objective of the rules is to create and reward displays of roses at the stage and with characteristics that are as close as possible to their most beautiful. Of course, all entries do not satisfy these criteria equally well, and that makes judging necessary. Yet rules and guidelines for judging beauty beg the question, What *is* beauty?

An economics professor, John Ise, at the University of Kansas, in writing about the United States national parks wrote words to this effect: "Crater Lake is the most beautiful lake in the world because no other lake can be so beautiful." When I first read those words, I thought they were just clever way of calling attention to an unusually attractive natural resource. I now believe John Ise had a concept of what beauty is that I have only recently come to understand. I believe that anything beautiful must satisfy one of two requirements:

- *It is sufficiently unusual to stimulate the imagination, or*
- *It is sufficiently perfect to command admiration.*

John Ise believed Crater Lake met both requirements. To him, Crater Lake was so perfect he could not imagine another lake more perfect. Both requirements are met by the classic beauty items with which I am familiar. The following dictionary descriptions of beauty capture both requirements:

- *The characteristic value of a beautiful thing apart from any effect it produces.*
- *The absolute perfection of the ideal or idea as suggested by, or reflected in, the relative sensuous perfection of works of art.*
- *The ideal itself apprehended through the medium of a beautiful thing.*

When I became a judge, I learned that I needed to know as much as I could learn about the characteristics of all rose types, not just hybrid teas and floribundas, types I had grown and shown. Roses can provide the maximum stimulation to the imagination and sense of the perfect only if the diversity of form and color of all rose types within the genus *Rosa* is considered.

AN ANALOGY

I have found it useful to relate beauty in roses to beautiful music. Consider a beautiful hybrid tea bloom on a long stem, enhanced with abundant, healthy foliage, as analogous to a vocal or an instrumental solo. A collection of (say) three roses becomes a trio in music. A bouquet or (say) a collection of 12 can be likened to a choir, and a rose garden may be likened to a symphony. Rose gardens have not been discussed to this point, and some elaboration is appropriate.

For many years, I thought of rose gardens as just a place to park rose bushes I wanted to grow. At some point, I began to consider the impression created by the rose garden taken as a whole. It became helpful to think of a prospective rose garden as a potential symphony— "a harmonious complexity or variety." I have come to see outdoor collections of rose bushes either as "parking lots" or rose gardens. I have created some "parking lots" I prefer to not identify. The following gardens reflect a "harmonious complexity," and I am proud of my association with them.

- *The Corvallis Rose Garden at Avery Park.* This garden was begun in 1955–56 and reflects the cooperative efforts of members of the Corvallis Rose Society and employees of the City of Corvallis since that time. It is a display garden accredited by All-America Rose Selections, a nonprofit association. All rose types may be found here. Giant sequoias impart a sense of grandeur. The garden is being made more accessible to those who are handicapped or disabled.

- *The Rose Garden at the Resources and Conservation Center, Washington, D. C.* The location is 1616 P Street Northwest, between O and P Streets in one direction and 16th and 17th in another. The rose garden is within a courtyard surrounded by attractive buildings. It was named in my honor in 2002. It is not far from 1600 Pennsylvania Avenue, where another rose garden may be found.

- *The First United Methodist Church Rose Garden in Corvallis, Oregon.* I have cared for this rose garden since 1986. Most major rose types may be found here. It exemplifies "harmonious variety," especially in the spring and fall, when all rose types are in bloom.

- *The Betty and Emery Castle Residence in Corvallis.* I planned this rose garden to enhance the Stoneybrook Retirement Village landscaping and the residence where it is located. Most rose types are displayed. It is easily accessible. I have cared for this garden since 1986 and have selected most of the roses that are in the garden at present.

Appendix C

A BRIEF DIGRESSION ON THOMAS KUHN'S
THE STRUCTURE OF SCIENTIFIC REVOLUTIONS

WHAT IS A DISCIPLINARY MATRIX?

Chapter 4 and Chapter 6, especially, include frequent references to the seminal work of Thomas Kuhn on scientific revolutions. The following paragraphs attempt to describe and illustrate that part of Kuhn's model that is used elsewhere in this book.

Thomas Kuhn taught that scientific revolutions and their aftermaths are not simple phenomena. A scientific revolution, according to Kuhn, is far more complex than a single flash of genius or a dramatic experiment that overthrows existing beliefs and thoughts. There are preconditions that must occur. One precondition is the existence of some part of reality that existing theory does not explain or rationalize well, but that might reasonably be expected of the existing theory. For example, the widespread unemployment that existed during the Great Depression years of my childhood could not easily be explained by the established and dominant economic theory at that time. Kuhn refers to these unexplained phenomena as anomalies. A revolutionary development will not only explain some or all of the anomalies unexplained by existing theory, it will also explain at least some of the reality covered by existing theory.

The previous paragraph is accurate as far as it goes, but it is incomplete. How does the layperson know if a different approach has replaced existing theory? Kuhn established a practical test. An idea is revolutionary, according to Kuhn, if it is sufficiently interesting to induce other scientists to elaborate the new approach. It is important to note that "revolutions" are a relative concept. They can be large or small, major or minor. One measure of the importance of a revolutionary idea is the number of scientific workers who are willing to devote some or all of their effort to its elaboration.

What does the term "elaborate" mean as I have used it here? Kuhn called the elaboration of a revolutionary discovery "normal science." I note three aspects of normal scientific effort. One pertains to *empirical* information made interesting to others, who are attracted to a field because they are excited about the new way of looking at things made possible by the revolutionary development. But the collection of factual information may not be a simple matter because new techniques may be required for the collection and measurement of the new, interesting information. A second important part of normal science, then, involves *instrumentation*, or the technology that supports normal science investigations. A third part of normal science is what Kuhn refers to as *puzzle-solving*. This involves exploration of inherent, but implicit, relations within the revolutionary advance. Answers to puzzles may, or may not, have practical relevance, but they are made interesting by the revolutionary advance.

Think now about the total enterprise that results from this combination of a revolutionary idea and the normal science activities that elaborate it. First is the collection and analysis of empirical information, often requiring developments in instrumentation technology, followed by the efforts to solve puzzles that arise from the revolutionary idea and the new empirical information that has been gathered. Colleges and universities develop graduate programs to involve students in existing research and to provide future personnel for research. Textbooks and literature collections may come into being and scientific journals requiring peer review may emerge to provide validation of the literature. It is this collection of activities that Kuhn describes as a disciplinary matrix. The content of a disciplinary matrix falls into two major categories. One category comprises concepts, principles, and subject-matter knowledge; the other is the infrastructure needed for the support of the other category. One reason that university disciplinary departments have been successful instruments of scientific research and graduate education is that both parts of Kuhn's disciplinary matrix may be found under one roof. I make repeated use of Kuhn's disciplinary matrix concept in this book.

The above descriptions of scientific revolutions and disciplinary matrixes is aided by Figure 1 in Chapter 3, which classifies certain notable economists, in particular John Maynard Keynes, whose name appears below that of Alfred Marshall. Keynes was an Englishman who studied economics at Cambridge University under Marshall. Prior to the revolution in economics brought about largely by Keynes, the structure of economics is best described as neoclassical microeconomics. This means that economic decisions were assumed to be made at the margin where demand and supply functions intersected to establish equilibrium conditions. It also means that these decisions were made mainly by individual producers and consumers in decentralized markets. Competition was assumed to prevail, although monopoly was recognized as possible. Under such circumstances, markets were expected to clear, and it was anticipated that essentially full employment would prevail.

John Maynard Keynes was a recognized and respected economist and a civil servant at the highest levels. In that capacity, he was concerned with both international and domestic affairs during and after World War I until World War II. During the Great Depression, he observed that unemployment was both high and persistent in, but not limited to, the economies of both Great Britain and the United States. To him, this was an anomaly to be attributed to the prevailing economic theory of that time—neoclassical microeconomics. Rather than viewing economics as consisting of individual decisions of producers and consumers in decentralized markets, he sought to explain the performance of entire economies from a centralized rather than a decentralized viewpoint. He sought to provide a theory of how an underemployment equilibrium could persist for a long period in market-oriented economies. His revolutionary book, *The General Theory of Employment, Interest and Money,* was published in 1936. To obtain some appreciation of the significance of the revolution that occurred in economics, one need only note that, prior to Keynes, the term microeconomics was not used to describe economics. The Keynesian approach was labeled "macro," and so "micro" became necessary to describe pre-Keynesian economics.

A normal science of macroeconomics appeared, as Kuhn would lead us to expect. As the title of Keynes's book makes explicit, macroeconomics directed attention to economy-wide variables. Different empirical information was made interesting by the revolutionary development— Keynesian economics. For example, the United States Department of Commerce created a new publication entitled *The Survey of Current Business,* which published information consistent with the new theory. Some economists became converts almost immediately. Others were not convinced by Keynes and tackled the puzzle of persistent unemployment from different perspectives. Macroeconomics textbooks appeared, and graduate students began to specialize in macroeconomics. Puzzles cried out for solutions, different empirical information was sought, and particular measurement techniques were required to accommodate macro variables.

Keynes recognized that knowledge accumulation in economics is an evolutionary process. He wrote that he "stood on the shoulders of giants." No doubt Marshall was one of those giants. And, as Keynes made clear, he was influenced by the classical economists, including Thomas Malthus. Within a few years, another revolutionary development occurred. By making use of mathematics, Paul Samuelson brought both macro- and microeconomics under the same neoclassical roof. Even so, the macro/micro distinction continues to exist.

NOTES AND REFERENCES

CHAPTER 1

1. There are interesting differences among the social science disciplines with respect to social capital. Coleman, a sociologist, noted that group arrangements often were needed to create an environment in which individual initiative can prosper. Political scientists have also been in the forefront of those writing about the subject (for example, R. D. Putnam). Elinor Ostrom, political scientist, has done original work on the effectiveness of group arrangements when numbers are limited and trust exists. Economists are all over the place in their considerations. Glenn Loury, an economist, is often cited as the first social scientist to employ social capital as an analytical tool. Joel Sobel has written a generally sympathetic literature review of social capital in which he discusses, and largely discounts, criticisms economists have levied against the concept. I, too, am sympathetic to greater use of the social capital concept (Castle 2002). If the criticisms that economists have levied against social capital were applied to human and natural capital, these concepts might fall into disuse as well.

The definition of social capital that follows next is the one I will use in this book and is the one I believe to be generally agreed upon in the social sciences. Social capital, then, exists when group arrangements are a more effective means of achieving individual ends than individual action individually determined. This definition does not specify how such arrangements are structured, but it assumes they are held together because a degree of trust exists among those constituting the group, which facilitates a mutual expectation of reciprocity. Alternatives to small-group action include individual and independent actions, as well as large-group control that may be economy- or nationwide in scope. A book by Tushaar Shah provides wonderful descriptions and examples of individual, small-group, and central decision making for irrigation development and management in South Asia.

I have difficulty understanding why such a common-sense description of something that obviously exists has created such a stir among the learned disciplines. Some economists apparently react negatively to social capital because it cannot be incorporated neatly into existing models. This may be a valid objection, but it may also reveal something about the completeness of such a theoretical structure. Other objections appear to stem from imagined political or ideological implications. An extreme conservative view may be suspicious of any type of group action that might be superior to individual effort. At the opposite extreme, there may be concern that small-group action necessarily will be at the expense of the aspirations and accomplishments of larger collective groups, such as states or nations. My view is that the most effective decision-making and action form is something to be discovered for problems and situations, rather than imposed a priori.

I have found the following forms of capital to be useful to be useful in analyzing conditions in the countryside:

- *Natural:* those parts of the natural environment of current or potential value to humans.

- *Human:* investments humans have made in themselves, individually or collectively, to enhance their capacity to satisfy human needs.

- *Human-created:* the hardware and software humans have created to enhance their productivity and satisfaction over time. A tractor, a roadway, and the contents of a library building constitute examples.

- *Social:* small-group arrangements that provide more effective means of achieving individual ends than either independent, individual action or central decision making by, say, a state or nation. A cooperative marketing group or a neighborhood association for the development and preservation of a local waterway provide examples of social capital.

2. Maureen Kilkenny and Monica Haddad have conducted research that is helpful in evaluation of one-room schools. They note that students generally benefit from mentoring and examples provided by peers. In large classes of students nearly all of the same age, the probability is great that for any given student there will be a classmate who can serve as such a peer. This, of course, is less true for one-room country schools with sparse enrollment. However, all grades are to be found there, and all in one room. The role that older children play as mentors and peers may have a greater probability of occurring in such an environment than in many others.

The consolidation of rural schools has long been a significant political issue in rural America. Many educators and others have argued that rural schools are often too small for effective education to be provided. It has also been argued that education cost per student can be reduced with consolidation. Both arguments have merit in some circumstances, but David Reynolds has noted that school consolidation has economic overtones for community development. The rural town or city that captures a consolidated school is assured of survival at least for a period.

The separation of education costs among parents, school districts, and state and federal government in most rural places is exceedingly complex and seldom has been related to quality of education. This is perhaps the most under-researched public policy issue concerning rural America.

Castle, Emery N. 1991. "The Benefits of Space and the Cost of Distance." Chapter 1 in *The Future of Rural America: Anticipating Policies for Constructive Change,* Kenneth Pigg, ed. Boulder, CO: Westview Press.

Castle, Emery N. 2002. "Social Capital: An Interdisciplinary Concept." *Rural Sociology* 67(3):331–49.

Coleman, James Samuel. 1988. "Social Capital in the Creation of Human Capital." *American Journal of Sociology* 94:S95–S120.

Egan, Timothy. 2006. *The Worst Hard Times.* Boston: Houghton-Mifflin.

Kilkenny, Maureen, and Monica Haddad. 2008. "Rural Human Capital Development." Chapter 9 in *Frontiers in Resource and Rural Economics: Human-Nature, Rural-Urban Interdependencies,* JunJie Wu, Paul W. Barkley, and Bruce A. Weber, eds.. Washington, DC: Resources for the Future.

Loury, Glenn. 1977. "A Dynamic Theory of Racial Income Differences." In *Women, Minorities, and Employment Discrimination,* P. Wallace and A. Lamond, eds. Lexington, MA: Lexington Books.

Ostrom, Elinor. 1990. *Governing the Commons: The Evolution of Institutions for Collective Action.* New York: Cambridge University Press.

Putnam, R. D. 2000. *Bowling Alone: The Collapse and Revival of American Community.* New York: Simon and Schuster.

Reynolds, David R. 1995. "Rural Education: Decentering the Consolidation Debate." Chapter 24 in *The Changing American Countryside: Rural People and Places,* Emery N. Castle, ed. Lawrence, KS: University Press of Kansas.

Shah, Tushaar. 2009. *Taming the Anarchy: Groundwater Governance in South Asia.* Washington, DC, and Colombo, Sri Lanka: Resources and International Water Management Institute.

Smith, Adam. 1776. *An Inquiry into the Nature and Causes of the Wealth of Nations.* New York: Modern Library, Random House (1994).

Sobel, Joel. 2002. "Can We Trust Social Capital?" *Journal of Economic Literature* 40:139–54.

United States Department of Agriculture. *1941 Yearbook of Agriculture: Climate and Man.* Washington, DC: Government Printing Office.

CHAPTER 2

1. I was the first of their four sons to enter the armed services. My two older brothers, Robert and Carl, were married and had children at that time. Don, my brother six years younger than I, was too young to be involved as a soldier in World War II. Robert and Carl did enter the service later in WW II but did

not leave the United States. Don was in the Navy during the Korean War. After he completed college, he became a commissioned officer in the Navy and served in the Korean theater.

2. Until I was assigned as a member of a B-17 crew, I never changed military locations together with anyone I knew. Nevertheless, I made friends readily and never felt lonely or isolated when I was in the armed services, except for the days and weeks of my second Jefferson Barracks assignment.

3. After my account of assembly in and above the clouds was written, a 2008 article by Leslie A. Lennox was published and came to my attention. Lennox was a pilot in the 95th Bomb Group and flew bombing missions in late 1944 and 1945. The 95th, 100th, and 390th Bomb Groups together constituted the 13th Wing. Undoubtedly, Lennox and I were on many of the same missions. I quote from two passages in his article:

> Each group had a pattern for the airplanes to fly during climb to assembly altitude. Some would fly in a triangle, some a rectangle and our group flew a circle . . . The patterns for each group fit together like a jigsaw puzzle. Unfortunately, strong winds aloft would destroy the integrity of these patterns, and there would be considerable over-running of each other's patterns . . . It was not uncommon to experience one or two near misses while climbing through the clouds, although you would never see the other airplane. You knew you had just had a near miss, when suddenly the airplane would shake violently as it hit the prop wash of another plane [p. 6].

4. Dates flown and mission (target):

10/15/44	Cologne**
10/17/44	Cologne**
10/19/44	Ludwigshafen & Mannheim
10/26/44	Hanover**
10/30/44	Merseburg (recalled, weather)
11/02/44	Merseburg**
12/24/44	Babenhausen, Kaiserlautern
12/25/44	Kaiserlautern**
12/31/44	Hamburg**
1/6/45	Annweiler
1/10/45	Cologne
1/14/45	Derben**
1/28/45	Duisburg
2/03/45	Berlin* **
2/06/45	Chemnitz
2/21/45	Nuremberg
2/24/45	Bremen**

2/26/45	Berlin**
3/03/45	Brunswick**
3/10/45	Dortmund
3/11/45	Hamburg
3/14/45	Hanover
3/17/45	Plauen
3/19/45	Jena
3/22/45	Alhorn
3/28/45	Hanover
3/31/45	Bad Berka
4/09/45	Munich
4/15/45	Royan
4/17/45	Aussig

*On February 3, 1945, the 100th Bomb Group led the Third Air Division to Berlin.
**A diary account of these 11 missions may be found in Sion's *Through Blue Skies to Hell*.

5. In his review of this book prior to publication, my colleague Steve Buccola wrote, "In one sense, namely under the assumption that humans follow their moral imperative—their true selves—war is never justified. But given that they may never follow that imperative completely, war often is justified. And I think war is actually quite risky for politicians. George Bush II is a living example. FDR would be vilified today, by both victor and vanquished, if we had lost World War II."

6. When I was in England flying combat missions, I wrote Merab Weber and asked if she would correspond with me. She agreed to do so, and we exchanged letters frequently until I was discharged. At that time, I went to see her. Merab had left teaching to work at Boeing Aircraft at Wichita, Kansas, during World War II.

Crosby, Harry, Jan Riddling, and Michael Faley. "History of the 100th Bomb Group (H)." http://www.100thbg.com/mainmenus/history/historysummary_home.htm.

Faley, Michael P. 2005. 100th Bomb Group Historian (correspondence, December 19).

Flatley, Teresa K. 1997. "Bizarre B-17 Collision Over the North Sea." *World War II* (May): 42–48.

Galbraith, John Kenneth. 1958. *The Affluent Society.* Cambridge, MA: The Riverside Press, pp. 161, 162.

Grant, Rebecca. 2008. "The Long Arm of the US Strategic Bombing Survey." *Air Force* 91(February): 2.

Grilley, Robert. 2003. *Return from Berlin: The Eye of a Navigator.* Madison, WI: University of Wisconsin Press.

Lennox, Leslie A. 2008. "The Mighty Eighth." *The Newsletter of the Air Force Navigators Observers Association* (July).

Levine, Alan J. 1992. *The Strategic Bombing of Germany, 1940–1945.* Westport, CT: Praeger Publishers, pp. 1, 2.

Rooney, Andy. 1995. *My War.* New York: Random House.

Sion, Edward M. 2007. *Through Blue Skies to Hell: America's 'Bloody 100th' in the Air War Over Germany.* Philadelphia: Newbury.

United States Strategic Bombing Survey. 1947. *Defeat of the German Air Force.* Washington: Government Printing Office.

Wilkins, Bill. 2005. Book Review of *Return from Berlin: The Eye of a Navigator* by Robert Grilley. In *The Air Force Navigators Association* 21(July): 3.

CHAPTER 3

1. Years later, I gave the presidential address to the American Agricultural Economics Association. I asked Professor Bagley to provide an introduction, and he agreed. This was most appropriate because his classes were responsible for my becoming a professional economist. As I write this, I'm looking at an autographed photograph of Ed Bagley that hangs on the wall of my study. Ed Bagley and I independently developed a love affair with roses. When we each learned of the other's interest in roses, our professional bond of economics was cemented by a common hobby—rose growing.

2. Our daughter, Cheryl Diana, was born while I was working for the masters degree. I was delighted to have a daughter because I had no sisters. I have never regretted it turned out this way. My sole regret is that Merab and I could not give Cheryl Diana either the biological or adopted sibling she wanted so very much. Cheryl, in common with many professors' kids, came to learn a great deal about the professions her parents chose, as well as becoming acquainted with our associates. The reason was that family vacations were often combined with travel to professional meetings. When Cheryl was in the eighth grade, she and Merab went with me on a sabbatical I took at Purdue. While there, we spent a week in New York where, among other delights, we saw *The Sound of Music* on Broadway.

3. Any account of graduate education in economics at Iowa State in those years would be seriously deficient if it did not provide some detail about Gerhard Tintner. Tintner was one of the early econometricians. He described econometrics as one-third economics, one-third statistics, and one-third mathematics. I had two courses for credit from him and audited another of his classes. Tintner had been raised and educated in Austria, but his English was better than he let on. We had not met before I took his first class. My dad died just as the fall quarter was getting underway, and I missed the first month in the first course I had from him. When I first went to his class, he asked that I see him in his office after class. I did so. He was kind, but he just told me to study

hard and not miss any more classes.

His method of teaching consisted in large part of calling students to the blackboard and asking them to explain some complicated subject that was in the reading assignments. When a student went to the board, Tintner would get very close and then shout instructions and comments as the student attempted to answer a question or explain a complicated passage from the reading assignment. He did not send me to the board the quarter I lost my dad, but he made up for it the next quarter. In the second course I had from him, he sent me to the board as we were reviewing for the final examination. I did not do a really good job with a question he asked, and he shouted, "Mein Gott!!! How can you be so stupid?!! Take your seat!!!" And then the following exchange occurred.

> TINTNER: "Mr. Castle, were you in my class last quarter?"
> EMERY: "Yes."
> TINTNER: "Did I pass you?"
> EMERY: "Yes."
> TINTNER: "Mein Gott! I vill not make the same mistake twice!!!"

In another class taken that quarter, he asked me to go to the board and explain an equation from Hicks's *Value and Capital*. I did so successfully and was feeling good when Tintner posed his next question. He went to a different side of the equation from where I began my earlier analysis, posed a hypothetical change in a variable, and asked me to explain its effect. I reflected a moment and blurted what seemed to me intuitively obvious, "The answer will be the same." Tintner was furious and shouted, "Mein Gott! How can you be so stupid? Take your seat, and I vill explain." I slunk back to my seat, covered with embarrassment but puzzled by how I could have made such a major mistake. Tintner picked up a fresh piece of chalk, cleared his throat, and then exclaimed, "Now, I vill explain!" He paused. He started through his explanation again, but soon discovered it was not leading him where he wanted to go. He then said, "I vill start again." For a second time, his explanation took him in the direction of my intuitive answer. He paused again, replaced the chalk in the chalk tray, and solemnly announced: "Class Dismissed."

4. The purpose of this brief summary is to provide a frame of reference for the non-economist reader of this book who wishes to place in perspective the economists referred to in the text. The reader is encouraged to make extensive use of Figure 1. The material included herein draws heavily on notes for a two-hour lecture I gave to the Academy of Lifelong Learning (ALL) at Oregon State University on April 14, 2007. Had I not previously given this lecture, I doubt I would have had the audacity to offer this summary.

5. In 1968, the Central Bank of Sweden instituted "The Sveriges Riksbank Prize in Economic Science in Memory of Alfred Nobel." This was the origin of the Nobel Laureate award in economic science. The award is given by the Royal Swedish

Academy of Sciences according to the same principles as those for Nobel prizes that have been awarded since 1901. I count seven economists named in Figure 1 who are Noble Laureates. Many included in that figure were not eligible to become Nobel Laureate because they were deceased before the awards became fully operable. The necessary and sufficient requirement for inclusion in Figure 1 was a direct or indirect impact on my intellectual journey. Even so, I consider it to be a distinguished group to be offered without apology.

6. I first met Ken Boulding shortly after I went to Oregon State in the mid-1950s. I had been given responsibility for advancing the graduate program in agricultural economics and made a considerable effort to acquaint the graduate students with major figures within economics and agricultural economics. I learned that Boulding was spending some time in the Economics Department at the University of Washington. It occurred to me that he might be persuaded to come to Oregon State for a seminar with our graduate students. I telephoned there and was connected to him. I explained what I had in mind, and he immediately expressed interest. It turned out that he intended to spend a long weekend in Portland with a friend who was a professor at Reed College. Luckily for me, the friend was otherwise occupied on Saturday of that weekend. We agreed that I would come by his friend's house on that Saturday morning.

I then extended an invitation to Charles Friday, an OSU Economics Department faculty member, to accompany me. We picked Ken Boulding up about 8 a.m. We were in Corvallis by 10 a.m., and he gave a short talk to a small group of faculty at 11 a.m. Word spread across campus earlier in the week that he would be on campus, and he met with and spoke to an interdepartmental group of faculty after we had lunch. We then met with a group of graduate students at approximately 2 p.m. Ken Boulding was superb with them. He was both substantive and funny. They were in stitches most of the time, but it was always about an economics subject that he induced them to examine in an unorthodox way. We then returned him to his friend's house in Portland after taking him to dinner en route to Portland from Corvallis. It would be difficult to exaggerate the beneficial impact his visit had on the morale and orientation of our graduate students. I later came to know Ken Boulding much better, but I never forgot the generous spirit he displayed on that Saturday. Years later, when I was president of Resources for the Future, at a dinner hosted by the president of the American Academy for the Advancement of Science, Kenneth Boulding and I were named Fellows of the Academy. To share that occasion with Ken Boulding was one of the more memorable events of my career.

7. The preliminary examination in economic theory was in two parts. One part consisted of writing answers to several questions from the students' committee. This was followed by an oral exam of approximately two hours. I believed I had done well on both parts of the examination. When the oral exam was completed, I was asked to leave the room. After a while, Earl Heady came out and told me I had passed. The other professors then came out and offered congratulations. Earl Heady then asked if I would like to go to the Union for

coffee. Of course, I accepted.

When we arrived at the Union, Earl's graduate students, my buddies, were seated around a large oblong table hoping we would come by. They could hardly contain their curiosity as to how I had fared. We were no more than seated when one blurted: "How did Emery do?" Earl replied, "Oh, pretty well. He did not get all excited and forget what little he did know." I felt both good and bad.

8. One of the more valuable outcomes of my time at the Kansas City Fed was becoming acquainted with John "Jack" Edwards. Jack came to the Fed as a research assistant in agricultural economics. He had a masters degree in agricultural economics from the University of Nebraska. We worked together and became friends. He later went to the University of Chicago for his doctorate. We became colleagues again some years later at Oregon State, when we attracted him away from the University of Idaho.

9. Harry Truman returned to Independence, Missouri, after his service as President of the United States, when I was at the Federal Reserve Bank of Kansas City. He took an office on the 11th floor of our building while arranging more permanent quarters. He once had lunch with us in the bank's officers' dining room. This was exciting indeed because I was an admirer. I learned from him the distinction between the importance of the person holding a position and the importance of the position the person holds.

One Saturday morning I went to my office accompanied by our 5-year-old daughter, Cheryl. When leaving, we walked through an empty corridor to the ninth-floor elevators. I punched the elevator button, the door opened, and there was Harry Truman, all alone, wearing a big grin! As Cheryl and I stepped onto the elevator, I was overcome by confusion. How should I address a former President, let alone provide an introduction for our cute 5-year-old?

Not to worry; Harry, ignoring me, said to Cheryl, "I remember when my little girl was just your age."

Cheryl replied immediately, "And what was her name?"

Harry laughed and said, "Margaret."

Cheryl responded, "That's nice."

When we reached the first floor, they walked hand-in-hand off the elevator and to the sidewalk in front of our building, talking all the time. I trailed awkwardly behind, not taking part in their conversation, but proud our 5-year-old was an accomplished conversationalist.

Castle, Emery N. 1952. "Some Aspects of the Crop-Share Lease." *Land Economics* 28:177–79.

Castle, Emery N. 1954a. "Adapting Western Farms to Uncertain Prices and Yields." *Technical Bulletin* 75. Manhattan, KS: Kansas Agricultural Experiment Station.

Castle, Emery N. 1954b. "Flexibility and Diversification as a Means of Meeting Price and Yield Uncertainty in Western Kansas." *Journal of Farm Economics* 36:376–83.

Castle, Emery N. 1954c. "Adaptation of the Farm Firm in Western Kansas to Conditions of Uncertainty." *Journal of Finance 9:414*.

Chamberlin, Edward H. 1933. *The Theory of Monopolistic Competition: A Re-orientation of the Theory of Value*. Cambridge: Harvard University Press.

Eatwell, John, Marray Milgate, and Peter Newman, eds. 1987. *The New Palgrave Dictionary of Economics*. London: Macmillan.

George, Henry. 1879. *Progress and Poverty*. New York: Doubleday & Company.

Gray, Alexander. 1931. *The Development of Economic Doctrine: An Introductory Survey*. London: Longmans, Green and Company.

Hicks, John R. 1939. *Value and Capital*. Oxford: Clarendon Press.

Keynes, John M. 1936. *The General Theory of Employment, Interest and Money*. London: Macmillian.

Knight, Frank H. 1921. *Risk, Uncertainty and Profit*. Boston: Houghton Mifflin.

Rosenberg, A. 1992. *Economics: Mathematical Politics or Science of Diminishing Returns?* Chicago: University of Chicago Press.

Samuelson, Paul. 1947. *Foundations of Economic Analysis*. Cambridge: Harvard University Press.

Smith, Adam. 1776. *The Wealth of Nations*. New York: Random House (later edition with an introduction by Max Lerner, 1937).

Viner, Jacob. 1928. "Adam Smith and Laissez-faire." In *Adam Smith 1776– 1926*, J. M. Clark et al., eds. Chicago: University of Chicago Press.

Warsh, David. 2006. *Knowledge and the Wealth of Nations: A Story of Economic Discovery*. New York: W. W. Norton & Company.

Wikipedia. "Kenneth Arrow." http://en.wikipedia.org/wiki/Kenneth_Arrow (accessed January 29, 2009).

Young, Allyn A. 1928. "Increasing Returns and Economic Progress." *Economic Journal* 38:527–42.

CHAPTER 4

1. In 2000, Charles Howe edited a special issue of the journal *Water Resources Research*, for which he had earlier served as editor. Several people were asked to discuss water issues as they were during the time the water committee on which I served was in existence. This special issue of *Water Resources Research* provides a sense of the intellectual excitement surrounding public policies related to water at that time (see Castle 2000 and related articles).

2. By 1962, I was eligible for sabbatical leave at Oregon State. An opportunity

arose for me to become a visiting professor at Purdue University for one semester. Merab, Cheryl, and I followed a massive January storm from Corvallis to West Lafayette, Indiana. I taught a course at Purdue and interacted extensively with faculty and graduate students. Two students working for the masters returned to OSU with me and obtained their doctorates from OSU. I formed deep and abiding friendships with several faculty, especially Lowell Hardin and G. Edward Schuh. Lowell was then department head and invited me to give my impressions of the department there in a farewell seminar. Theirs was a much larger department than the one at OSU, and they appeared to have almost unlimited resources at their disposal. After I returned to OSU, Purdue made me a most attractive offer to return there on a permanent basis. I declined, but the Purdue experience permitted me to see my field of work in much broader perspective. My time at Purdue was an important part of my intellectual journey and opened my eyes to the potential of academic administration, due in large part to the wisdom of Lowell Hardin. The trip to Purdue, side trips, and the return to Corvallis were great family experiences. During spring break we spent a week in New York City.

3. Shortly after my appointment as dean of faculty, a reporter for the OSU student newspaper, *The Barometer,* interviewed President Jensen. The student asked Jensen if it was "risky" to place a person without administrative experience in the dean of faculty position. As I recall, Jensen replied along the following lines: "All first-time appointments to campus-wide positions are risky to some extent. Three qualities are essential for success campus-wide—hard work, analytical ability, and courage. Castle has demonstrated he works hard and has analytical ability. We will know about courage at this level only by trial. The same would be true for academic deans and department heads. In this respect, the Castle appointment is no more risky than any others I might make."

4. Just as I became dean of faculty, it was necessary that Oregon State respond to an unprecedented challenge. John Mosser, an Oregon legislator, persuaded the Legislature to pass, and Governor Mark Hatfield to sign, a bill that would make $1,000 awards to "good teachers" in the Oregon State System of Higher Education if students assisted in their selection. Mosser and others contended that higher education did not provide appropriate financial incentive for the encouragement of good teaching. Faculty were concerned. This was a clear intrusion into the traditional autonomy of universities in faculty evaluation. Yet if a faculty were to refuse to make such awards, it might suggest they feared student opinion. At Oregon State, the matter was debated in Faculty Senate. When a vote was taken, a majority favored making the awards. The responsibility was logically assigned to the dean of faculty office to select the (as I recall) 64 faculty who would receive $1,000 awards. I was fortunate that Oregon State had a long-standing program in support of good teaching. A teaching evaluation form was in existence that had been designed by students and faculty. We required that nominations for these awards would be made by students as supported by teaching evaluation forms completed by students.

Joint student-faculty recommendations for awards were made to my office. When the awardees were announced, there appeared to be a consensus across campus that many good teachers had been selected. By that time, everyone appeared to be weary of the entire adventure, and, so far as I know, nothing similar has been attempted since. Of course, student evaluation of teachers has become common practice and influences salary and promotion decisions.

5. "Visioning," without reference to resource constraints and possible changes in the operating environment, often leads to unrealistic expectations. I know of two organizations that have been damaged seriously by misuse of this methodology.

6. The advancement of a new theory or philosophy does not necessarily mean that earlier theories or philosophies will disappear. Thus, the new approach may serve to increase the alternatives available. The personal characteristics of the scientist, such as risk aversion, may then play a greater role in the approach or method selected.

Brown, William G., Ajmer Singh, and Emery N. Castle. 1964. "An Economic Evaluation of the Oregon Salmon-Steelhead Sports Fishery." *Oregon Agricultural Experiment Station Technical Bulletin 78*. Corvallis, OR: Oregon State University.

Castle, Emery N., and Carroll Dwyer. 1956. "Irrigation Possibilities in the Fort Rock Area." *Oregon Agricultural Experiment Station Circular 558*. Corvallis, OR: Oregon State University.

Castle, Emery N., and Manning H. Becker. 1962. *Farm Business Management: The Decision-making Process*. New York: Macmillan (2nd edition with Frederick J. Smith, 1972; 3rd edition with Gene Nelson, 1987).

Castle, Emery N., M. Kelso, and D. Gardner. 1963. "Water Resource Development: A Review of the New Federal Evaluation Procedures." *Journal of Farm Economics* 45:693–704.

Castle, Emery N. 1965. "The Market Mechanism, Externalities and Land Economics." *Journal of Farm Economics* 47(August):542–46.

Castle, Emery N. 1970. "Priorities in Agricultural Economics for the 1970s." *American Journal of Agricultural Economics* 52:831–40.

Castle, Emery N., C. W. Hovland, and J. G. Knudsen. 1970. *Report to the President of Oregon State University from the Commission on University Goals*. Corvallis, OR: Oregon State University.

Castle, Emery N. 1971. "The University in Contemporary Society." *American Journal of Agricultural Economics* 53:551–56.

Castle, Emery N. 2000. "Reflections on Water Policy and Scholarship in Economics." *Water Resources Update* 67(3):331–49.

Castle, Emery N. 2008. "Frontiers in Resource and Rural Economics: A Methodological Perspective." Chapter 12 in *Frontiers in Resource and Rural Economics,* JunJie Wu, Paul Barkley, and Bruce Weber, eds. Washington, DC: Resources for the Future.

Ciriacy-Wantrup, S. V. 1952. *Resource Conservation, Economics and Policies.* Berkeley: University of California Press.

Clawson, Marion. 1981. "Methods of Measuring the Demand for the Value of Outdoor Recreation." *Reprint 10.* Washington, DC: Resources for the Future.

Kuhn, Thomas S. 1996. *The Structure of Scientific Revolutions,* 3rd ed. Chicago: University of Chicago Press.

Nelson, Michael. 1973. *The Development of Tropical Lands: Policy Issues in Latin America.* Baltimore, MD: Johns Hopkins University Press.

Regan, Mark M. 1964. "Sharing Financial Responsibility of River Basin Development." Chapter 13 in *Economics and Public Policy in Water Resource Development,* Stephen C. Smith and Emery N. Castle, eds. Ames, IA: Iowa State University Press.

Stevens, Joe B. 1965. *A Study of Conflict in Natural Resource Use: Evaluation of recreational benefits as related to changes in water quality.* Doctoral dissertation. Corvallis, OR: Oregon State University.

Stevens, Joe B. 1966. "Recreational Benefits from Water Pollution Control." *Water Resources Research* 2:167–82.

Stoevener, Herbert H., J. B. Stevens, H. F. Horton, A. Sokoloski, L. P. Parrish, and E. N. Castle. 1972. "Multi-disciplinary Study of Water Quality Relationships: A Case Study of Yaquina Bay, Oregon." *Oregon Agricultural Experiment Station Special Reprint 348.* Corvallis, OR: Oregon State University.

U.S. Interagency Committee on Water Resources. 1958. *Proposed Practices for Economic Analysis of River Basin Projects.* Report to the Interagency Committee on Evaluation Standards, rev. ed.

CHAPTER 5

1. Shortly after I went to RFF, I invited S. V. Ciriacy-Wantrup, resource economist at the University of California, to RFF as a member of an advisory group. While there, he and Hitch came together. They shook hands, and Charlie Hitch said, "I do not recall seeing much of you when I was at Berkeley." Wantrup replied, "That is correct. You were always in that administration building." Upon hearing this exchange, I reflected, but not aloud: "The hermit and the recluse have met."

2. McGeorge Bundy, of course, was Eastern establishment through and through. He had served President John Kennedy as his national security advisor, been provost of Harvard, and received his education at Harvard and Yale.

3. When Merab and I were discussing whether to accept the RFF offer of employment, she asked, "How stable and safe is Resources for the Future?"

I replied, "RFF has just been given a four-year grant by the Ford Foundation. Clearly they have the support of the Ford Foundation. The Ford Foundation has to qualify as one of the safest institutions around." I was not greatly concerned about my own welfare as a result of the Marshall Robinson message, but I really did not relish telling Merab about it. The news would make my earlier assessment of the Ford Foundation to her dead wrong, as it affected us. I did, however, tell her soon after the Robinson visit. She listened carefully and then dismissed the matter with a statement along the following lines, "It appears you might have to get another job if things don't work out as we thought they would."

4. When Elwood Davis invited me to the Alfalfa Club dinner on one occasion, he said his wife would like the pleasure of Merab's company that evening. He further said that "Davy" Jones's wife would be with them. "Davy" Jones was General David Jones, the Chairman of the Joint Chiefs of Staff for the Armed Forces of the United States. General Omar Bradley was in town, and the military wives were honoring his wife that evening. Elwood Davis said arrangements had been made for Merab to be taken to her home at the end of the evening, permitting us to return home separately. When I returned home after my Alfalfa Club dinner, I found Merab had been there only a short time. I asked if she had a good time, and she replied, "Oh, yes, but I had a close call." I asked with some concern, "What do you mean?" She said, "When it was time to go home, it was explained to us we should go just outside to a particular place and we would be picked up by a limousine. They also said that if there were not enough limousines, taxis would be available." Merab then said, "For a while there, I thought I would have to take a taxi home." I said to her, "Well! It has not taken you long to become addicted to special privilege!" She agreed.

5. Would I have been heir apparent if the RFF board had not faced a near-crisis situation? The answer to this question can never be known. It was clear I had the near-unanimous, perhaps unanimous, support of the RFF staff. It is probable a majority of the board would have favored my appointment. Nevertheless, high-level administrators in two major foundations did not believe I was the best person to provide RFF leadership. So far as I could ascertain, their concerns about me were: 1) I was not well known in certain circles, 2) my education was not from prestigious institutions, and 3) I did not have a charismatic personality.

6. Merab and I lived in a motel on the edge of the OSU campus when we returned to Corvallis after our decade in Washington, D.C. The day after our return, I walked to the campus intending to go to my new office there. After I had walked a short distance on campus, I encountered a former associate. He greeted me cordially and we chatted for a while. He asked if I were on my way to my office, and I replied that I was indeed intending to go there. As we talked, I came to realize he did not know I had been away from OSU for about 10 years. I was amused, but thought little about it because he was senior to me in age.

When I entered my new office, I went immediately to my telephone and dialed the number of John Ahearne, RFF vice president. After the telephone there rang several times, a voice I did not recognize said, "Resources for the Future, may I help you?" I replied, "I wish to speak with John Ahearne." When the operator said, "Dr. Ahearne is not here today," I asked if I might leave a message, and the operator said: "Yes, you may. Whom shall I say is calling?" I said, "Emery Castle." To which she responded, "Would you spell that, please?"

I then mused to myself: "The people at OSU do not know I have been away. The people at RFF do not know I have been there. Perhaps the RFF experience was just a figment of my imagination."

Caldwell, Bruce J. 1982. *Beyond Positivism: Economic Methodology in the Twentieth Century*. New York: Routledge (rev. ed. 1984).

Castle, Emery N. 1965. "The Market Mechanism, Externalities and Land Economics." *Journal of Farm Economics* 47(August):542–56.

Castle, Emery N. 1990. "Resources for the Future as a Policy Research Institute: Organizational Characteristics and Performance, 1952–1986." Corvallis, OR: University Graduate Faculty of Economics, Oregon State University.

Castle, Emery N. 2008. "Frontiers In Resource and Rural Economics: A Methodological Perspective." Chapter 12 in *Frontiers in Resource and Rural Economics: Human-Nature, Rural-Urban Interdependencies*, JunJie Wu, Paul W. Barkley, and Bruce A. Weber, eds. Washington, DC: Resources for the Future.

Hall, R. J., and C. J. Hitch. 1939. "Price Theory and Business Behavior." *Oxford Economic Papers* May(2):22–25.

Hausman, D. I. 1992. *The Inexact and Separate Science of Economics*. Cambridge: Cambridge University Press.

Jarrett, Henry, and Irving Fox, with a foreword by Charles Hitch. 1977. *Resources for the Future: The First Twenty-Five Years*. Washington, DC: Resources for the Future.

Krutilla, John. 1967. "Conservation Reconsidered." *American Economic Review* 51(777):86.

McCloskey, D. 1983. "The Rhetoric of Economics." *Journal of Economic Literature* 21:481–517.

Samuelson, Paul. 1955. "Diagrammatic Exposition of a Theory of Public Expenditures." *Review of Economics and Statistics* 36 (November):387–89.

CHAPTER 6

1. I arranged with Dr. Russell Youmans, director of the Western Rural Development Center, then located at Oregon State, to have an office in his center. I had employed Russ for Extension work when he first came to OSU in 1966,

and we had been friends from that time. The four regional Rural Development Centers, a part of the Land Grant–USDA systems, became co-operators of the National Rural Studies Committee when it came into existence in 1987. I am indebted to Russ Youmans for his assistance in providing infrastructure in support of my return to Oregon State in 1986.

2. Maureen Kilkenny credits a National Rural Studies Committee article for attracting her to this subject (Castle 1993b).

3. One day early in the 21st century, Bill Boggess, then head of the OSU Department of Agricultural and Resource Economics, invited me and Rich Adams, professor of Agricultural and Resource Economics, to lunch. He wished to discuss with us the possible loss of JunJie Wu, a young professor in his department, to the University of Wisconsin, which had extended an offer to JunJie. After we had discussed several options, I authorized Bill to tell JunJie that an endowed professorship would become available in the OSU Department of Agricultural and Resource Economics upon my demise. Bill, of course, knew I had established this with the OSU Foundation, but he also knew I did not intend to make a public announcement of it during my lifetime.

Bill immediately seized upon this and soon articulated plans to establish the possibility of an endowed chair after sufficient funds were obtained for a professorship. Bill asked if I were willing for the professorship to be awarded to JunJie if sufficient funds were available. I thought highly of JunJie but told Bill he would have to be selected in accordance with the guidelines for the professorship that had been established with the Foundation. In summary, Bill was successful in persuading JunJie to remain at OSU, sufficient pledges and funds were obtained to endow the professorship (and nearly enough to endow a chair), and, in accordance with the guidelines I had established, JunJie was awarded the professorship, which he now occupies.

In 2005, a symposium was held in my honor to explore intellectual frontiers in resource and rural economics. JunJie chaired the symposium, with Bruce Weber and Paul Barkley as organizing committee members. The three then became editors of a book that resulted from the symposium (Wu, Weber, Barkley 2005).

I have not been privileged to have many children, although I treasure greatly my one biological daughter and my numerous stepchildren, step-grandchildren, and step-greatgrandchildren. Nevertheless, I have had numerous graduate students and younger professional associates during the course of my career. To observe, and participate in, their professional growth and development has been, for me, the source of enormous joy and pride. The brilliant JunJie Wu, reared in rural China and occupying the professorship that carries my name, has become a treasured friend and colleague.

4. The NRSC met near Greenfield, Mississippi, for its second annual meeting. This area is one of the most poverty-stricken rural places in the nation. One day we had lunch with a group of local African-Americans assembled by the Cooperative Extension Service. During lunch, I sat beside a middle-aged

woman who said she was a single parent for "both of her families." That meant that she was parent of three adult children who were no longer members of her household, as well as two younger children who were still at home. We had sufficient time for a conversation that permitted us to discuss several subjects in depth.

She said she had encouraged her older children to prepare themselves so that they could have a better life than she'd had. As adults, they were able to find employment elsewhere, and, in her words, were "doing alright." One lived in Texas, and, as I recall, two were in Illinois. She told me when she was working to prepare them for a better life as they were growing to adulthood, she did not appreciate that she was preparing them to leave her community and move to places where she could not see them regularly. As a consequence, she said she became depressed and overweight. She added she then understood that it was inevitable that her children would leave her community because there were few opportunities there. She also said she wanted her two younger children to obtain as much education as possible.

I then asked her if she had taken part in any of the civil rights marches in Mississippi. She replied, "No, but I have marched." For what cause," I asked. She then said: "When my children got to the fourth grade, I thought they should have books."

5. For a demonstration of the necessary and sufficient conditions for appropriate intermediate decision making in a particular location, see Castle, Chapter 12, in Wu, Barkley, and Weber, eds., *Frontiers in Resource and Rural Economics.* Washington, DC: Resources for the Future.

6. In my judgment, the most powerful objection to an integrated path was not articulated until after my report to John Byrne had been submitted. When reviewing this book, Steven Buccola wrote: "Integrating outreach with on-campus research also creates a problem: it widens the cultural gap within the department, making it difficult to arrive at shared understandings of performance standards and realizations. In a word, it works against, even as it potentially enriches, community. This is, as I see it, the Land Grant University Problem." Buccola makes a most basic point. It is my belief that the problem he describes will become less severe with the passage of time although it is unlikely to disappear.

7. I marvel at my good fortune. Betty Rose (Thompson) Burchfield and I embarked upon the 21st century hand in hand, married in a modest ceremony in Portland, Oregon, in the year 2000. We both grew up in the Great Plains—South Dakota for her, Kansas for me—during those Drought and Depression years. We share many, many of the same values.

Our paths nearly crossed several times prior to our actual meeting, which occurred sometime after 1990. She was a civilian employee at the Sioux Falls Army Air Base after I did my tour of duty there. She then went to Washington, D. C., and worked in the Pentagon during World War II. She worked for the Air Force; I served in it. While at the Pentagon, she typed and delivered the morning report for General Hap Arnold, Commanding General, Army Air Force.

In 1954 we both moved to Oregon with our respective families. She was the mother of a son, Rick, at that time seven years of age. My daughter, Cheryl, was a six-year-old when Merab and I came to Oregon. They lived in Seaside, and then Ontario, Oregon. We were Corvallis settlers.

In 1986, both families moved to the same neighborhood in Corvallis. For different reasons, both couples had selected Corvallis as a place to retire. Not long after Betty and her husband, Mason, settled, he developed brain cancer and died in 1990. My wife, Merab, was diagnosed with Alzheimer's disease shortly after we moved to Corvallis. Betty befriended Merab some time before Betty and I met.

Betty has a large family, and I like and admire the whole bunch. Her five siblings all live in Wessington Springs, South Dakota, and have children, grandchildren, and great grandchildren galore. Betty has three children, five grandchildren, and three great-grandchildren. Betty and my daughter, Cheryl, have a great relationship.

Betty is so self-sufficient I find it difficult to find things I can do for her that she cannot do better for herself. An exception occurred in 2002. RFF celebrated its 50th anniversary that year, and we went to Washington, D.C., for some of the events. We attended an RFF dinner where Gwen Ifill was the featured speaker. Betty was seated at one of the head tables with the RFF President and Vice President. I was seated at the other beside Gwen Ifill. It had just been announced that Gwen was to become moderator of *Washington Week in Review*. I knew Betty was envious of me because Gwen Ifill was a heroine to her. During dinner I told Gwen that the smallish white-haired lady at the next table thought decision makers at National Public Television were brilliant, lucky, or both to name her as moderator of *Washington Week in Review*. After Gwen's presentation, and as the group was disbanding, I noted Gwen and Betty were in deep conversation. I did something for Betty better than she could do for herself.

As this is written, Betty Rose and I are in the 10th year of our marriage. Each helps the other adjust to old age with particular attention to emotional needs, health, nutrition, and exercise.

I marvel, repeatedly, at my good fortune.

Arthur, W. Brian. 1994. *Increasing Returns and Path Dependence in the Economy*. Ann Arbor, MI: University of Michigan Press.

Backhouse, Roger E., and Jeff Biddle, eds. 2000. *Toward a History of Applied Economics*. Durham and London: Duke University Press.

Castle, Emery N. 1985. "The Forgotten Hinterlands: Rural America." President's Essay in 1985 *Annual Report*. Washington, DC: Resources for the Future.

Castle, Emery N. 1991."The Benefits of Space and the Cost of Distance."
Chapter 1 in *The Future of Rural America: Anticipating Policies for Constructive Change,* Kenneth Pigg, ed. Boulder, CO: Westview Press.

Castle, Emery N. 1993a. "Rural Diversity: An American Asset." *The Annals of the American Society of Political and Social Science* (September).

Castle, Emery N. 1993b. "On the University's Third Mission: Extended Education." Corvallis, OR: Oregon State University. (Final report to President John V. Byrne on the placement of the Oregon State University Extension Service within the university.)

Castle, Emery N. 1997. *Final Report: National Rural Studies Committee.* Corvallis, OR: Western Rural Development Center, Oregon State University.

Castle, Emery N. 1998. "A Conceptual Framework for the Study of Rural Places." *American Journal of Agricultural Economics* 80(3): 621–31.

Castle, Emery N. 2002. "Social Capital: An Interdisciplinary Concept." *Rural Sociology* 67(3) September.

Castle, Emery N., and David Ervin. 2008. "Frontiers in Resource and Rural Economics: A Synthesis." Chapter 1 in *Frontiers in Resource and Rural Economics,* JunJie Wu, Paul W. Barkley, and Bruce A. Weber, eds. Washington, DC: Resources for the Future.

Castle, Emery N., and Bruce A. Weber. 2006. "Policy and Place: Requirements of a Space-Based Policy." Working Paper AREC O6-O1, RSP 06-01. Corvallis, OR: Oregon State University.

Castle, Emery N. 2008. "Frontiers in Rural and Resource Economics: A Methodological Perspective." Chapter 12 in *Frontiers in Resource and Rural Economics,* JunJie Wu, Paul W. Barkley, and Bruce A. Weber, eds. Washington, DC: Resources for the Future.

Crosson, Pierre. 1995. "The Use and Management of Rural Space." Chapter 7 in *The Changing American Countryside,* E. Castle, ed. Lawrence, KS: University Press of Kansas.

Kilkinney, Maureen. 2008. "The New Rural Economics." Chapter 5 in *Frontiers in Resource and Rural Economics,* JunJie Wu, Paul W. Barkley, and Bruce A. Weber, eds. Washington, DC: Resources for the Future.

Kuhn, Thomas S. 1996. *The Structure of Scientific Revolutions,* 3rd ed. Chicago: University of Chicago Press.

Kunce, Mitch, and Jason F. Shogren. 2008. "Property Taxation and the Redistribution of Rural Resource Rents." Chapter 10 in *Frontiers in Resource and Rural Economics,* JunJie Wu, Paul W. Barkley, and Bruce A. Weber, eds. Washington, DC: Resources for the Future.

Lewis, Peirce. 1995. "The Urban Invasion of Rural America: The Emergence of the Galactic City." Chapter 3 in *The Changing American Countryside: Rural People and Places,* E. N. Castle, ed. Lawrence: The University Press of Kansas.

Mills, Edwin S. 1995. "The Location of Economic Activity in Rural and Nonmetropolitan United States." Chapter 6 in *The Changing American Countryside*, E. N. Castle, ed. Lawrence, KS: University Press of Kansas, pp. 103–33.

Oakerson, Ronald J. 1995. "Structures and Patterns of Rural Governance." Chapter 21 in *The Changing American Countryside: Rural People and Places*, E. N. Castle, ed. Lawrence: The University Press of Kansas.

Oakerson, Ronald J. 2008. "The Politics of Place." Chapter 11 in *Frontiers in Resource and Rural Economics*, JunJie Wu, Paul W. Barkley, and Bruce A. Weber, eds. Washington, DC: Resources for the Future.

Ostrom, Elinor. 1990. *The Evolution of Institutions for Collective Action.* Cambridge: Cambridge University Press.

Segerson, Kathleen. 2008. "Resources and Rural Communities: Looking Ahead." Chapter 13 in *Frontiers in Resource and Rural Economics*, JunJie Wu, Paul W. Barkley, and Bruce A. Weber, eds. Washington, DC: Resources for the Future.

Starrs, Paul F. 1995. "Conflict and Change on the Landscapes of the Arid American West." Chapter 14 in *The Changing American Countryside: Rural People and Places*, E. N. Castle, ed. Lawrence: The University Press of Kansas.

Summers, Gene F. 1995. "Persistent Rural Poverty." Chapter 11 in *The Changing American Countryside*, E. Castle, ed. Lawrence: University of Kansas, pp. 213–28.

Trow, Martin. 1979. "Reflections on the Transition from Mass to Universal Higher Education." *Daedalus* (winter):1–42.

Weber, Bruce A. 2008. "People and Places at the Ragged Edge." Chapter 8 in *Frontiers in Resource and Rural Economics*, JunJie Wu, Paul W. Barkley, and Bruce A. Weber, eds. Washington, DC: Resources for the Future.

CHAPTER 7

1. The word *externalities* used in this book is derived from economic theory. An externality occurs when an economic decision inflicts a gain or loss on others in the economic system that is not reflected in economic system prices. In other words, externalities are created by a production or consumption decision that imposes a gain or loss on others in the economic system by a means other than prices (Varian 1984).

2. My former student Dr. Sandra Batie has written a recent article on "wicked" problems. There is substantial overlap between her "wicked" and my "messy" problems. We discuss overlapping issues from different perspectives.

3. In Chapter 4, research on the economics of water quality in Yaquina Bay, Oregon, is described. Given the state of the art at that time, this was indeed a messy problem. Had the conventional wisdom of the time (the 1960s) in resource economics been applied, a pulp and paper mill would have been recommended

for the estuary. This would have done major damage to the aquatic resources that supported a minor recreational industry. The investigation we conducted was designed to identify possible decisions that would permit the mill to operate and also protect the natural resources of the estuary. It was discovered there were numerous options that would permit such an outcome (multiple equilibria). A solution was forged by the local community, state government, and Georgia Pacific Corporation, made possible by the natural comparative advantage of the estuary. The local community had need for a stable payroll provided by the mill. It also wished to protect its natural resources, which would have been harmed greatly by unregulated discharge of the effluent into the estuary.

The Yaquina Bay study in the 1960s was a complex problem given the state of the art of resource economics at that time. Bruce Weber and I have since developed a more general theoretical model for analyzing complex community decision-making issues. This model permits path dependence, social capital, and multiple equilibria to be taken into account (Arthur 1994). In 2008 I applied this model to the Yaquina Bay economy retrospectively. Instead of a traditional expansion path as defined in economic theory, the trajectory used was influenced greatly by path dependence in most years. From time to time, exogenous shocks have occurred, and the trajectory shifted, although some degree of path dependence has remained. The pulp and paper mill continues to operate, but its relative importance has declined. Conversely, the natural resources of the estuary have increased in economic importance. The Oregon State University Hatfield Marine Science Center has become established there, and a commercial aquarium has located there. The messy problem posed by Yaquina Bay has yielded to the tools of complexity economics, although they were not known to us by that label when they were employed.

Arthur, W. Brian. 1994. *Increasing Returns and Path Dependence in the Economy.* Ann Arbor: University of Michigan Press.

Batie, Sandra S. 2008. "Wicked Problems and Applied Economics." *American Journal of Agricultural Economics* 90(5):1176–91.

Brady, Nyle C., ed.1967. *Agriculture and the Quality of Our Environment.* Washington, DC: American Association for the Advancement of Science Publication 85.

Castle, Emery N., Maurice Kelso, and Delworth Gardner. 1963. "Water Resource Development: A review of the new federal evaluation procedures." *Journal of Farm Economics* 45:693–704.

Castle, Emery N. 1967. "Economic and Administrative Problems of Water Pollution." In *Agriculture and the Quality of Our Environment,* N. C. Brady, ed. Washington, D.C.: American Association for the Advancement of Science Publication 85:251–85.

Castle, Emery N., C. Warren Hovland, and James G. Knudsen. 1970. *Report to the President of Oregon State University from the Commission on University Goals.* Corvallis, OR: Oregon State University.

Castle, Emery N. 1981. "Agricultural Education and Research: Academic Crown Jewels or Country Cousin?" *The 1980 Kellogg Foundation Lecture to the National Association of State Universities and Land Grant Colleges.* Washington, DC: Resources for the Future. (Also published, exclusive of appendixes, in *Proceedings of the 94th Annual Meeting of the National Association of State Universities and Land Grant Colleges, November 1980,* held in Atlanta, Georgia.)

Castle, Emery N. 2008. "Frontiers in Resource and Rural Economics: A Methodological Perspective." Chapter 12 in *Frontiers in Resource and Rural Economics,* JunJie Wu, Paul W. Barkley, and Bruce A. Weber, eds. Washington, DC: Resources for the Future.

Evenson, Robert E., Paul E. Waggoner, and Vernon W. Ruttan. 1979. "Economic Benefits from Research: An Example from Agriculture." *Science* (September 14)1101–07.

Kuhn, Thomas S. 1996. *The Structure of Scientific Revolutions,* 3rd ed. Chicago: University of Chicago Press.

Varian, Hal R. 1984. *Microeconomic Analysis,* 2nd ed. New York, London: W. W. Norton.

CHAPTER 8

1. Formal study typically is not used to describe the reading one does in preparation for teaching. In this case, I believe it is appropriate to refer to what I did as formal. It was in a defined field and required sufficient mastery to explain and discuss it with others. Knowledge was acquired in a disciplined and structured way. We likely would describe the preparation a student makes for class as formal study. Is it incorrect to do the same for teachers?

After I began to hold administrative positions, I often thought of my after-work study of methodology and economic theory as my "Go to Hell" effort. Part of my motivation for such study was to avoid becoming so subject-matter obsolete that administration would be my only professional opportunity. I wanted always to be unafraid to say: "Go to Hell!"

2. When I was at RFF, my wife and I were invited to dinner one evening in the District of Columbia. A neighbor of our host and hostess and his wife were also there, as were a few others. The neighbor was David Acheson, son of Dean Acheson, who was Secretary of State for part of the time Harry Truman was President. It was a small group, and there was abundant time for good conversation before and during dinner.

David Acheson discovered I knew something about our nation's history when his father was Secretary of State, and he welcomed the opportunity to converse with someone who did. At one point, I said I had long been curious how two people with such different backgrounds as were his father's and Harry Truman's had become close associates, obviously based on mutual respect and admiration. David Acheson replied: "*What you have said is true. They did*

have different backgrounds, but the answer to your question is simple. Neither ever forgot who was President."
Abraham Lincoln never forgot he was President of the United States.

3. I have before me a statement that reads, "In 2007 the largest 50% of farms produced 99.6% of the total farm gate value [of agricultural commodities]. This means the smallest 50% produced 0.4%." If an attempt were made to measure the economic value of the gross domestic product contributed by institutions serving the smallest 50% of farmers, it certainly would not be a large number. Yet if those institutions were measured by the number of farms served, the contribution would be equal in importance to those serving the largest 50 percent of farms. I rest my case that the importance of nonmarket rural institutions should not be measured by their contribution to the quantities of goods and services valued in the market place (Castle 2005, 196).

4. In 2003, I was invited by the Northeastern Agricultural and Resource Economics Association to give a paper that was subsequently published in its *Agricultural and Resource Economics Review* [32/1(April 2003):18-32]. A revision was published in 2006 as Chapter 1 in an RFF book titled *Economics and Contemporary Land Use Policy: Development at the Rural-Urban Fringe*, edited by Richard J. Johnston and Stephen K. Swallow. The journal article has the title "Land, Economic Change, and Agricultural Economics." The book chapter carries the title "Land, Economic Change, and Economic Doctrine."

In addition to the change in title, the article had an appendix labeled "Intellectual Constructs and the Limits of their Applicability: A Digression" that was not appended to the book chapter. This appendix discusses financial instruments and the Enron and Long Term Capital Management failures. This applied agricultural, resource, rural economist, with no particular special expertise in macro economics, was concerned in 2003 about the danger to our economy created by permitting newly discovered financial instruments to become our master rather than be our servant. The appendix was not included with the book chapter. I suspect the book editors eliminated the appendix in part because they wanted to spare me the embarrassment of writing about macroeconomics in a book on land. If so, they were correct. It did not belong there, but it belonged someplace, written by someone with greater expertise in macro and financial economics.

Arrow, K., P. Dasgupta, L. Goulder, G. Daily, P. Ehrlich, G. Heal, S. Levin, K-G. Maler, S. Schneider, D. Starrett, and B. Walker. 2004. "Are We Consuming Too Much?" *Journal of Economic Perspectives* 18(4):147–72.

Castle, Emery N., and Bruce A. Weber. 2006. *Policy and Place: Requirements of a Successful Place-Based Policy.* Working Paper RSP 06-01. Corvallis, OR: Department of Agricultural Economics, Oregon State University.

Castle, Emery N. 2008. "Frontiers in Resource and Rural Economics: A Methodological Perspective." Chapter 12 in *Frontiers in Rural and*

Resource Eonomics, JunJie Wu, Paul W. Barkley, and Bruce A. Weber, eds. Washington, DC: Resources for the Future.

Lewis, Peirce. 1995. "The Urban Invasion of Rural America: The Emergence of the Galactic City." Chapter 3 in *The Changing American Countryside: Rural People and Places,* E. N. Castle, ed. Lawrence: The University Press of Kansas.

Segerson, Kathleen. 2008. "Resources and Rural Communities: Looking Ahead." Chapter 13 in *Frontiers in Resource and Rural Economics,* JunJie Wu, Paul W. Barkley, and Bruce A. Weber, eds. Washington, D.C.: Resources for the Future.

Veblen, Thorstein. 1898. "Why Is Economics not an Evolutionary Science?" *Quarterly Journal of Economics,* July.

Weber, Bruce A. 2007. "Rural Poverty: Why Should States Care and What Can State Policy Do?" *Journal of Regional Analysis and Policy* -37(1):48–52.

World Commission on Environment and Development. 1987. *Our Common Future.* Oxford: Oxford University Press.

APPENDIX A

1. Cheryl Castle Delozier lost her husband of 26 years (C. Robert "Bob" Delozier) to cancer in October 2008. She and Rodney Rogers were married in 2010.

APPENDIX B

Castle, Emery N. 1999. *The Memorial Rose Garden at the First United Methodist Church.* Corvallis, OR: First United Methodist Church.

Ise, John. 1961. *Our National Park Policy: A Critical History.* Washington, DC: Published for Resources for the Future by Johns Hopkins University Press.

Webster's Third New International Dictionary. 1981. Springfield, MA: Merriam-Webster, Inc.

INDEX

The initials ENC indicate Emery N. Castle.